P.O.263203 71-451-6/9/71

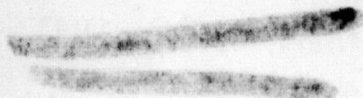

SEQUENCES AND COMBINATORIAL PROBLEMS

THE POCKET MATHEMATICAL LIBRARY
Jacob T. Schwartz, *Editor*

PRIMERS:

1. THE COORDINATE METHOD
by I. M. Gelfand et al.

2. FUNCTIONS AND GRAPHS
by I. M. Gelfand et al.

WORKBOOKS:

1. SEQUENCES AND COMBINATORIAL PROBLEMS:
by S. I. Gelfand et al.

2. LEARN LIMITS THROUGH PROBLEMS!
by S. I. Gelfand et al.

3. MATHEMATICAL PROBLEMS: AN ANTHOLOGY
by E. B. Dynkin et al.

COURSES:

1. LIMITS AND CONTINUITY
by P. P. Korovkin

2. DIFFERENTIATION
by P. P. Korovkin

SEQUENCES AND
COMBINATORIAL PROBLEMS

BY

S. I. GELFAND, M. L. GERVER, A. A. KIRILLOV,
N. N. KONSTANTINOV, A. G. KUSHNIRENKO

Revised English Edition
Translated and Freely Adapted by
RICHARD A. SILVERMAN

GORDON AND BREACH

SCIENCE PUBLISHERS

NEW YORK · LONDON · PARIS

Contents

Preface

This book has a very simple structure. It begins with a brief section called "Preliminaries" presenting the modicum of background information needed to solve the 89 problems stated in the next section, called "Sample Problems." These problems are, for the most part, equipped with hints or answers or both. But the nub of the book is the section called "Detailed Solutions," where you will find all 89 sample problems worked out in full detail. In our opinion, just studying these solutions (after first spending a decent amount of time trying to solve the problems on your own!) is a perfectly plausible way of learning about sequences and combinatorial problems. Finally, to make sure you have mastered the subject matter of the book, you should attack all 37 problems in the section called "Test Problems." In fact, think of this section as a (rather tough) final examination on which you must get at least a passing grade. Good luck!

Preliminaries

1. Sequences

Suppose a real number u_n is associated with every positive integer n. Then the numbers

$$u_1, u_2, \ldots, u_n, \ldots$$

are said to form an *infinite sequence* or simply a *sequence*, and the numbers themselves are called the *terms* of the sequence. The following are all examples of sequences:

1. The sequence all of whose terms are ones:

$$1, 1, \ldots, 1, \ldots$$

Here $u_n = 1$, i.e., the number 1 is associated with every positive integer.

2. The sequence of consecutive odd numbers:

$$1, 3, 5, \ldots$$

Here $u_n = 2n - 1$.

3. Consider the sequence specified by the formula

$$u_n = \frac{n(n+1)}{2}.$$

Write out the first seven terms of the sequence, and find u_{100}, u_{n-3} and u_{n+1}.

1

Answer. 1, 3, 6, 10, 15, 21, 28, ...

$$u_{100} = 5050,$$

$$u_{n-3} = \frac{(n-3)(n-3+1)}{2} = \frac{(n-3)(n-2)}{2},$$

$$u_{n+1} = \frac{(n+1)(n+2)}{2}.$$

The *n*th term of a sequence is called the *general term* of the sequence. A sequence is often specified by giving an explicit formula for its general term. Thus the sequence such that $u_n = n^2$ starts off like

$$1, 4, 9, 16, 25, 36, 49, ...$$

The sequence with general term u_n is often denoted by $\{u_n\}$, i.e., by writing the general term inside curly brackets.

Example 1. Find a formula for the general term of the sequence

$$2, 5, 8, 11, 14, 17, 20, ...$$

Answer. One possibility is

$$u_n = 3n - 1.$$

Another is

$$u_n = (3n - 1) \times \text{(the number of digits in } n\text{)}.$$

There are an infinite number of other possibilities!

Example 2. As a less obvious problem, find a formula for the general term of the sequence

$$0, \frac{7}{2}, 13, \frac{63}{2}, 62, \frac{215}{2}, 171, ...$$

Answer. One possibility is

$$u_n = \frac{n^3 - 1}{2}.$$

Example 3. One could hardly write a reasonable formula for the general term of the sequence

$$3, 1, 4, 1, 5, 9, 2, 6, 5, 3, 5, \dots \quad (1)$$

Nevertheless, (1) is just the sequence whose nth term is the nth digit in the decimal representation of the number $\pi = 3.1415926\dots$ Hence there is actually a definite rule associating a term u_n of the sequence (1) with every positive integer n. For example, $u_1 = 3$, $u_2 = 1$, $u_7 = 2$, etc. Thus, despite the absence of an explicit formula for the general term of this sequence, it is possible, at least in principle, to find the number in any given position, be it the first, seventh or 1007th position. In particular, it can be shown that the sequence (1) is not periodic, i.e., that no block of terms repeats itself over and over again like the underlined digits in the decimal expansion of

$$\tfrac{1}{7} = 0.\underline{142857}\underline{142857}\dots$$

As you recall from elementary geometry, π is the ratio of the circumference of a circle to its diameter. The number π was first estimated by Archimedes, who knew that it lies between $3\tfrac{1}{7}$ and $3\tfrac{10}{71}$. For ordinary purposes, the value $\pi = 3.14$ is often good enough, but more accurate values like $\pi = 3.14159$ are sometimes needed. In principle, π can be computed to any desired accuracy. However, it is hard to imagine a problem requiring knowledge of π to more than ten decimal places or so. Believe it or not, somebody called Shanks once wasted a large part of his life calculating π to 707 places. Then, more recently, several thousand digits of π were calculated on a high-speed electronic computer, and it turned out that Shanks' calculations were incorrect starting from the 202nd digit!

Example 4. Calculate the first 10 digits of the sequence

$$1, 1, 2, 3, 5, \dots \quad (2)$$

formed by the following rule: The first two terms equal 1 ($u_1 = 1, u_2 = 1$), while starting with the third term, every term is the sum of the preceding two terms, i.e.,

$$u_n = u_{n-1} + u_{n-2} \quad (n \geqslant 3).$$

There is an explicit formula for u_n in this case, but it is not too simple (see Prob. 43, p. 15). The terms of the sequence (2) are called the *Fibonacci numbers*, and the sequence itself is called the *Fibonacci sequence*.

Example 5. Find the first few terms of the sequence

$$u_1, u_2, \ldots$$

whose nth term equals the sum of all the positive integers from 1 to n inclusive.

Answer. 1, 3, 6, 10, 15, 21, ...

Solution. $u_1 = 1, \quad u_2 = 1 + 2 = 3, \quad u_3 = 1 + 2 + 3 = 6,$
$u_4 = 1 + 2 + 3 + 4 = 10, \ldots, u_n = 1 + 2 + \cdots + n, \ldots$

2. Mathematical Induction

If the first person in a line is a woman and if there is another woman standing behind every woman (except the last), then every person in the line is a woman. The reasoning behind this somewhat facetious example occurs again and again in mathematics and is called the *principle of mathematical induction*. We now give a more serious formulation of this principle: *Given a sequence of assertions, if the first assertion is true and if every true assertion is followed by another true assertion, then every assertion in the sequence is true.*

Example 1. Prove that

$$1 + 2 + 3 + \cdots + n = \frac{n(n+1)}{2} \tag{3}$$

for every positive integer n. This formula comprises a whole sequence of assertions:

1)
$$1 = \frac{1 \cdot 2}{2},$$

2)
$$1 + 2 = \frac{2 \cdot 3}{2},$$

3)
$$1 + 2 + 3 = \frac{3 \cdot 4}{2},$$

4)
$$1 + 2 + 3 + 4 = \frac{4 \cdot 5}{2},$$

. . . .

The first assertion of obviously true. We now verify that every true assertion is followed by another true assertion. Suppose assertion k is true, i.e., suppose (3) is valid for $n = k$ so that

$$1 + 2 + 3 + \cdots + k = \frac{k(k + 1)}{2}. \qquad (4)$$

Adding $k + 1$ to both sides of (4), we obtain

$$1 + 2 + 3 + \cdots + k + (k + 1) = \frac{k(k + 1)}{2} + (k + 1)$$

$$= \frac{(k + 1)(k + 2)}{2}.$$

But this is just assertion $k + 1$, which comes right after assertion k. Thus we have shown that every true assertion is followed by another true assertion. Hence, according to the principle of mathematical induction, every assertion in the sequence is true, i.e., formula (3) holds for every positive integer n.

The same problem can be solved without recourse to mathematical induction. Writing

$$u_n = 1 + 2 + 3 + \cdots + n,$$

we have

$$u_n = 1 + 2 + 3 + \cdots + (n - 2) + (n - 1) + n, \qquad (5)$$

and

$$u_n = n + (n - 1) + (n - 2) + \cdots + 3 + 2 + 1 \qquad (6)$$

(the second sum is the first sum written backwards). Adding equations (5) and (6), we find that

$$2u_n = [1 + n] + [2 + (n - 1)] + [3 + (n - 2)] + \cdots$$
$$+ [(n - 2) + 3] + [(n - 1) + 2] + [n + 1].$$

Each term in brackets equals $n + 1$, and there are exactly n such terms. In other words,

$$2u_n = \underbrace{(n + 1) + (n + 1) + \cdots + (n + 1) + (n + 1)}_{n \text{ times}} = n(n + 1),$$

and hence

$$u_n = \frac{n(n + 1)}{2},$$

which is just another way of writing (3).

Another somewhat different form of the principle of mathematical induction goes as follows: *Given any assertion involving an arbitrary positive integer n, suppose that*
   a) *The assertion is true for $n = 1$;*
   b) *Validity of the assertion for $n = k$ implies its validity for $n = k + 1$.*
Then the assertion is true for every positive integer n.

   Warning. Note that hypothesis b) does not state that the assertion is true for $n = k + 1$ but only that *if* it is true for $n = k$, *then* it is true for $n = k + 1$.

   Example 2. Prove than $n^5 - n$ is divisible by 5 for every positive integer n. The proof involves two steps:
   a) If $n = 1$, $n^5 - n$ equals 0 and hence is trivially divisible by 5.
   b) Let $n = k$ be an arbitrary positive integer k, and suppose $k^5 - k$ is divisible by 5. Then $(k + 1)^5 - (k + 1)$ is also

divisible by 5. In fact, it follows from

$$(k + 1)^5 = k^5 + 5k^4 + 10k^3 + 10k^2 + 5k + 1$$

that

$$(k + 1)^5 - (k + 1) = (k^5 + 5k^4 + 10k^3 + 10k^2 + 5k + 1)$$
$$= (k^5 - k) + 5 (k^4 + 2k^3 + 2k^2 + k).$$

But each of the terms on the right is divisible by 5, the first by hypothesis, the second since it is obviously a multiple of 5. Since the sum of two numbers divisible by 5 is itself divisible by 5, it follows that $(k + 1)^5 - (k + 1)$ is divisible by 5. Thus hypotheses a) and b) figuring in the second formulation of the principle of mathematical induction are satisfied. Therefore $n^5 - n$ is divisible by 5 for every positive integer n.

There are other versions of the principle of mathematical induction equivalent to those just given. For example, *given any assertion involving an arbitrary positive integer n, suppose that*

a) *The assertion is true for* $n = 1$;

b) *Validity of the assertion for* $n \leqslant k$ *implies its validity for* $n = k + 1$.

Then the assertion is true for every positive integer n.

The difference between this formulation of the principle of mathematical induction and the preceding one is that hypothesis b) now states that the assertion is true for all $n \leqslant k$ and not just for $n = k$. However, it is easy to see that the two formulations are equivalent in the sense that every theorem which can be proved by applying the principle of mathematical induction in one form can also be proved by applying the principle in the other form (think this through).

It should also be noted that hypothesis a) is only needed to "get started," so to speak. Suppose that hypothesis a) is replaced by a new hypothesis

a′) *The assertion is true for* $n = 8$, say,

while hypothesis b) is left the same. Then it follows from a′) and b) that our assertion is true for all n starting from 8. This is precisely the situation encountered in Problem 9, p. 9.

Sample Problems

We begin with a group of problems (1–14) involving the principle of mathematical induction.

1. Prove that

$$1^3 + 2^3 + 3^3 + \cdots + n^3 = \frac{n^2(n+1)^2}{4}$$

for every positive integer n.

2. Prove that

$$\frac{1}{1 \cdot 2} + \frac{1}{2 \cdot 3} + \cdots + \frac{1}{n(n+1)} = \frac{n}{n+1}$$

for every positive integer n.

Hint. The problem can be solved in two ways, either by mathematical induction or by using the formula

$$\frac{1}{n(n+1)} = \frac{1}{n} - \frac{1}{n+1}.$$

3. Calculate the sum

$$\frac{1}{1 \cdot 4} + \frac{1}{4 \cdot 7} + \cdots + \frac{1}{(3n-2)(3n+1)}.$$

Answer. $\dfrac{n}{3n+1}$.

4. Prove that

$$\frac{1}{a(a+1)} + \frac{1}{(a+1)(a+2)} + \cdots + \frac{1}{(a+n-1)(a+n)}$$

$$= \frac{n}{a(a+n)}$$

8

holds for every positive integer n and every a not equal to zero or a negative integer.

5. Prove that

$$1 \cdot 1! + 2 \cdot 2! + \cdots + n \cdot n! = (n + 1)! - 1,$$

where n is any positive integer and $n!$ (read "n factorial") denotes the product of all positive integers from 1 to n inclusive.

6. Prove that

$$\left(1 - \frac{1}{4}\right)\left(1 - \frac{1}{9}\right) \cdots \left(1 - \frac{1}{n^2}\right) = \frac{n + 1}{2n}$$

for every integer $n \geqslant 2$.

Hint. The problem can be solved in two ways, either by mathematical induction or by using the formula

$$1 - \frac{1}{n^2} = \frac{(n - 1)(n + 1)}{n^2}.$$

In using mathematical induction, start from $n = 2$.

7. Verify that

$$1 - 2^2 + 3^2 - \cdots - (-1)^{n-1} n^2 = (-1)^{n-1} \frac{n(n + 1)}{2}.$$

8. Into how many parts is the plane divided by n straight lines no two of which are parallel and no three of which intersect in a single point?

Hint. Suppose n lines have already been drawn, and show that drawing the $(n + 1)$st line increases the number of parts of the plane by $n + 1$.

Answer. $\dfrac{n^2 + n + 2}{2}$.

9. Prove that any integer greater than 7 can be written as a sum consisting of the integers 3 and 5 exclusively. (For example, $8 = 3 + 5, 9 = 3 + 3 + 3, 10 = 5 + 5, 11 = 3 + 3 + 5$, etc.)

Hint. Given that $k (k > 7)$ can be written as a sum of threes and fives, show that $k + 1$ can also be written as a sum of threes

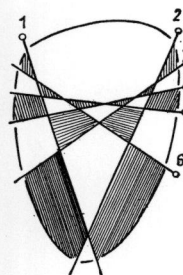

Fig. 1

and fives. Then apply mathematical induction, starting from $n = 8$.

10. Suppose the plane is derived into parts by n straight lines as in Problem 8. Prove that the plane can be colored black and white in such a way that any two parts with a common side have different colors, as shown in Figure 1 for the case $n = 6$. (Such a coloring is said to be *regular*.)

11. Prove that the sum of the cubes of three consecutive positive integers is divisible by 9.

12. Verify that
$$11^{n+2} + 12^{2n+1}$$

is divisible by 133 for any integer $n \geqslant 0$.

13. Prove that the inequality

$$(1 + a)^n > 1 + na$$

holds for any integer $n \geqslant 2$ provided that $a > -1$ and $a \neq 0$.

14. Prove that

$$1 + \frac{1}{\sqrt{2}} + \frac{1}{\sqrt{3}} + \cdots + \frac{1}{\sqrt{n}} > \sqrt{n}$$

for any integer $n \geqslant 2$.

Hint. Use the inequality

$$\frac{1}{\sqrt{k + 1}} > \sqrt{k + 1} - \sqrt{k}.$$

The next group of problems (15–43) deals with sequences.

15. Consider the sequence $u_1, u_2, \ldots$ such that

$$u_1 = 2, \quad u_2 = 3, \quad u_n = 3u_{n-1} - 2u_{n-2} \quad (n \geqslant 2).$$

Prove that
$$u_n = 2^{n-1} + 1.$$

16. Consider the sequence $u_1, u_2, \ldots$ such that

$$u_1 = 1, \quad u_{n+1} = u_n + 8n \quad (n \geqslant 1).$$

Prove that $u_n = (2n - 1)^2$.

17. Given a sequence $u_1, u_2, \ldots$, let $\Delta u_1, \Delta u_2, \ldots$ be the sequence with general term

$$\Delta u_n = u_{n+1} - u_n$$

(the capital Greek letter Δ is read "delta"). This new sequence is called the sequence of *first differences* of the original sequence $u_1, u_2, \ldots$. Given that $u_n = n^2$, what is Δu_n?

Answer. $\Delta u_n = 2n + 1$.

18. Suppose two sequences $u_1, u_2, \ldots$ and $v_1, v_2, \ldots$ have the same first differences, i.e., suppose $\Delta u_n = \Delta v_n$ for all n.

a) Can it be asserted that $u_n = v_n$?

b) Can it be asserted that $u_n = v_n$ provided it is known that $u_1 = v_1$?

Answer. a) No; b) Yes. Use the formula $u_1 + \Delta u_1 + \Delta u_2 + \cdots + \Delta u_{n-1} = u_n$.

19. Suppose every term of as equence $w_1, w_2, \ldots$ equals the sum of the corresponding terms of the sequences $u_1, u_2, \ldots$ and $v_1, v_2, \ldots$, i.e., suppose $w_n = u_n + v_n$ for all n. Prove that $\Delta w_n = \Delta u_n + \Delta v_n$ for all n.

20. Given a sequence with general term $u_n = n^k$, prove that the general term Δu_n of the sequence of first differences is a polynomial of degree $k - 1$ in n. Find the leading coefficient of this polynomial.

Answer. The leading coefficient is k.

21. Given a sequence whose general term u_n is a polynomial of degree k in n with leading coefficient a_0, prove that Δu_n is a polynomial of degree $k - 1$ in n. Find the leading coefficient of this polynomial.

Hint. Use the results of Problems 19 and 20.

22. Starting from a sequence

$$u_1, u_2, \ldots, u_n, \ldots,$$

from the sequence of first differences

$$\Delta u_1 = u_2 - u_1, \ \Delta u_2 = u_3 - u_2, \ldots, \Delta u_n = u_{n+1} - u_n, \ldots,$$

then the sequence of *second differences*

$$\Delta^2 u_1 = \Delta u_2 - \Delta u_1,\ \Delta^2 u_2 = \Delta u_3 - \Delta u_2, \ldots,$$
$$\Delta^2 u_n = \Delta u_{n+1} - \Delta u_n, \ldots$$

and so on, k times up to the sequence of *kth differences*

$$\Delta^k u_1 = \Delta^{k-1} u_2 - \Delta^{k-1} u_1,\ \Delta^k u_2 = \Delta^{k-1} u_3 - \Delta^{k-1} u_2, \ldots,$$
$$\Delta^k u_n = \Delta^{k-1} u_{n+1} - \Delta^{k-1} u_n, \ldots$$

$u_n = n^3$

n	n^3	First difference	Second difference	Third difference
1	1	7		
2	8	19	12	6
3	27	37	18	6
4	64	61	24	6
5	125	91	30	⋮
6	216	⋮	⋮	
⋮	⋮			

Fig. 2

For example, if $u_n = n^3$, then the first, second and third differences are as shown in Figure 2. Suppose $u_n = n^k$. Show that all the terms of the sequence of kth differences equal the same number. What is this number?

Answer. $k!$

23. Given a sequence $\{v_n\} = \{n^k\}$, prove that there is another sequence $\{u_n\}$ such that

a) $\Delta u_n = v_n$;

b) u_n is a polynomial of degree $k + 1$ in n.

What is the leading coefficient of this polynomial?

Hint. Use induction in k and the results of Problems 19 and 20.

24. Suppose a sequence $\{u_n\}$ is such that every term of the sequence of fourth differences $\{\Delta^4 u_n\}$ vanishes. Prove that u_n is a polynomial of degree 3 in n.

25. Given a sequence $\{u_n\}$ such that Δu_n is a polynomial of degree k in n, prove that u_n is a polynomial of degree $k + 1$ in n.

Hint. Use the results of Problems 19 and 23.

26. Calculate the sum

$$1^2 + 2^2 + \cdots + n^2.$$

Hint. Use Problem 25 to show that the sum is a polynomial of degree 3 in n.

Answer. $\dfrac{n(n+1)(2n+1)}{6}$.

27. Calculate the sum

$$1 \cdot 2 + 2 \cdot 3 + \cdots + n\,(n + 1).$$

Answer. $\dfrac{n\,(n + 1)\,(n + 2)}{3}$.

28. By an *arithmetic progression* is meant a sequence $\{u_n\}$ such that

$$u_{n+1} = u_n + d$$

for all n, where the number d is called the *difference* of the progression. By a *geometric progression* is meant a sequence $\{u_n\}$ such that

$$u_{n+1} = qu_n,$$

where the number q is called the *ratio* of the progression.

a) Given an arithmetic progression $\{u_n\}$, write a formula for u_n in terms of u_1 and the difference d.

b) Given a geometric progression $\{u_n\}$, write a formula for u_n in terms of u_1 and the ratio q.

Answer. a) $u_n = u_1 + (n - 1)\,d$; b) $u_n = q^{n-1}u_1$.

29. a) Find a formula for the sum

$$S_n = u_1 + u_2 + \cdots + u_n$$

of the first n terms of an arithmetic progression.

b) Find a formula for the product

$$P_n = u_1 u_2 \cdots u_n$$

of the first n terms of a geometric progression.

Answer. a) $S_n = nu_1 + \dfrac{n\,(n - 1)}{2}\,d$; b) $P_n = u_1^n q^{n(n-1)/2}$.

30. a) What is the sum of the first 15 terms of the arithmetic progression with first term 0 and difference $\frac{1}{3}$?

b) What is the product of the first 15 terms of the geometric progression with first term 1 and ratio $\sqrt[3]{10}$?

Answer. a) 35; b) 10^{35}.

31. a) The third term of an arithmetic progression equals 0. Find the sum of the first five terms.

b) The third term of a geometric progression equals 4. Find the product of the first five terms.

Answer. a) 0;   b) $4^5 = 1024$.

32. Let S_n be the sum of the first n terms of a geometric progression with first term u_1 and ratio q. Prove that

$$S_n = \frac{q^n - 1}{q - 1} u_1.$$

Hint. Consider the quantity $qS_n - S_n$.

33. A messenger arrives in a city of three million inhabitants, bearing interesting news which he tells to two other people 10 minutes later. Then 10 minuter later each of these people tells the news to two more people (who haven't heard the news yet) and so on, as long as there are people who don't know the news. How long does it take the whole city to learn the news?

Answer. 3 hours and 30 minutes.

34. A bicyclist and a horseback rider compete in a five-lap race. Both take the same time to complete the first lap, but the bicyclist traverses each remaining lap 1.1 times more slowly than the preceding lap. The horseback rider also slows down, but by the same amount on each remaining lap. Both contestants arrive at the finish line at exactly the same time. Who takes longer to complete the fifth lap, and how much longer does he take?

Answer. It takes the horseback rider 0.985 less time than the bicyclist to complete the fifth lap.

35. Find the sum S of all odd numbers less than 1000.

Answer. $S = 250,000$.

36. Find the sum S of all three-digit numbers not divisible by 2 or by 3.

Hint. Let $S^{(1)}$ denote the sum of all three-digit numbers, $S^{(2)}$

the sum of all three-digit numbers divisible by 2, $S^{(3)}$ the sum of all three-digit numbers divisible by 3 and $S^{(6)}$ the sum of all three-digit numbers divisible by 6. Then $S = S^{(1)} - S^{(2)} - S^{(3)} + S^{(6)}$.

Answer. $S = 164,700$.

37. Suppose the sum of the first n terms of a sequence equals $S_n = 3n^2$. Prove that the sequence is an arithmetic progression. Find the first term u_1 and the difference d of the progression.

Answer. $u_1 = 3$, $d = 6$.

38. Does there exist a geometric progression with the numbers 27, 8 and 12 as terms (these numbers need not appear consecutively or in the order given here)? At what positions in the sequence can these numbers appear?

Answer. Yes, such a progression exists. If 27 appears in the mth place, 8 in the nth place and 12 in the pth place, then m, n and p satisfy the equation $m - 3p + 2n = 0$.

39. Answer the same questions for the numbers 1, 2 and 5.

Answer. No such progression exists.

40. The squares of the twelfth, thirteenth and fifteenth terms of an arithmetic progression form a geometric progression. Find all possible values of the ratio of the geometric progression.

Answer. The ratio can take the values 1, 4, $\frac{13}{2} + \frac{3}{2}\sqrt{17}$, $\frac{13}{2} - \frac{3}{2}\sqrt{17}$.

41. Find all geometric progressions $\{u_n\}$ such that

$$u_n = u_{n-1} + u_{n-2}$$

for all $n \geqslant 3$.

Answer. Any progression with ratio $\frac{1}{2} + \frac{1}{2}\sqrt{5}$ or $\frac{1}{2} - \frac{1}{2}\sqrt{5}$.

42. The terms of a certain sequence are sums of the corresponding terms of two geometric progressions. What is the third term of the sequence?

Answer. The third term also equals zero.

43. Represent the terms of the Fibonacci sequence

$$w_1 = 1, w_2 = 1, w_n = w_{n-1} + w_{n-2} \quad (n \geqslant 3)$$

(see Example 4, p. 3) as sums of corresponding terms of two geometric progressions $\{u_n\}$ and $\{v_n\}$.

Answer. $\dfrac{1}{\sqrt{5}} \left[\left(\dfrac{1 + \sqrt{5}}{2} \right)^n - \left(\dfrac{1 - \sqrt{5}}{2} \right)^n \right]$.

Here is one way of lighting the bulbs:

Here is another (the bulbs are all off):

Fig. 3

The remaining problems (44–94) are all of a combinatorial character, i.e., they typically ask "how many ways are there of doing such and such?" Problems of this kind, involving the calculation of numbers of distinct combinations, are of great importance in probability theory, computational mathematics, the theory of automata and mathematical economics.

44. In how many ways can five light bulbs be lit?

Hint. Each bulb can be on or off. Two ways of lighting the bulbs are distinct if they differ in the state of at least one bulb (see Figures 3 and 4, where white and black indicate on and off respectively). There are various approaches to the problem. One is to simply enumerate all possible ways of lighting the bulbs, namely

These ways are different:

So are these:

Fig. 4

0) The bulbs are all off:

• • • • •   (1 way)

1) One bulb is on:

○ • • • •
• ○ • • •
• • ○ • •
• • • ○ •
• • • • ○   (5 ways)

2) Two bulbs are on (draw a figure and count the ways).

(? ways)

3) Three bulbs are on. (Don't hurry. There is no need to draw a separate figure or count. Think things over and you will find that you already know the answer.)   (? ways)

4) Four bulbs are on (verify that the number of ways is the same as in case 1).   (5 ways)

5) The bulbs are all on:

○ ○ ○ ○ ○   (1 way)

Another approach goes as follows: Suppose there is one bulb instead of five. Then how many ways can the bulb be lit? Next suppose there are two bulbs, three bulbs, etc. How many times does the number of ways of lighting the bulbs increase each time an extra bulb is included?

Answer. There are 32 ways of lighting five bulbs.

45. Given n light bulbs, let C_k^n denote the number of ways of turning on k bulbs $(k = 1, ..., n)$. Thus, in the preceding problem we found that $C_0^5 = 1$ while $C_1^5 = C_4^5 = 5$. Prove that

$$C_0^n + C_1^n + \cdots + C_k^n + \cdots + C_n^n = 2^n.$$

Hint. Using the symbol C_k^n allows us to express the mathematical content of the sentence "There are 10 ways of lighting 5 bulbs such that 3 bulbs are on" by the formula

$$C_3^5 = 10.$$

A comparison of the two methods of solving Problem 44 leads to the formula

$$C_0^5 + C_1^5 + C_2^5 + C_3^5 + C_4^5 + C_5^5 = 2^5.$$

46. A city has n traffic lights, each of which can be in any of three states (green, yellow or red). In how many ways can the traffic lights be lit?

Answer. 3^n.

47. How many ways can n traffic lights be lit if k lights can be in any of three states (green, yellow or red), while $n - k$ lights can be in either of two states (green or red)?

Answer. $3^k 2^{n-k}$.

48. What is the largest number of distinct license plates consisting of 3 letters followed by 4 digits?

49. How many six-digit numbers are there containing no zeros or eights?

Hint. Suppose you already know the number of five-digit numbers containing no zeros or eights. Now can you calculate the number of six-digit numbers with the same property?

Answer. $8^6 = 262,144$.

50. In a certain country, no two people have the same set of teeth, i.e., one person may have a full set of teeth, one may be missing the lower left molar, another the upper right molar, and so on. What is the maximum population of the country?

Hint. There are 32 teeth in a full set.

51. Imagine that the expression

$$(x - 1)(x - 2) \cdots (x - 100)$$

is multiplied out and all terms involving x to the same power are combined. What is the coefficient of x^{99}?

Answer. -5050.

52. How many ways are there of writing a positive integer $n \geqslant 2$ as a sum of two positive integers? There are two ways of interpreting this problem, i.e., with or without the assumption that the order of terms is important. In other words, it can be assumed that $8 = 3 + 5$ and $8 = 5 + 3$ are two ways of writing 8 as a sum of two terms or just one way. The answers differ in the two cases. Solve both problems.

Answer. There are $n - 1$ ways if the order is important. Otherwise there are $\frac{1}{2}n$ ways if n is even and $\frac{1}{2}(n - 1)$ ways if n is odd.

53. How many ways are there of writing a positive integer $n \geqslant 3$ as a sum of three positive integers if order is important?

Answer. $\dfrac{(n - 1)(n - 2)}{2}$.

54. By *Pascal's triangle* is meant the array

$$
\begin{array}{ccccccccc}
 & & & & 1 & & & & \\
 & & & 1 & & 1 & & & \\
 & & 1 & & 2 & & 1 & & \\
 & 1 & & 3 & & 3 & & 1 & \\
1 & & 4 & & 6 & & 4 & & 1 \\
1 & & 5 & & 10 & & 10 & & 5 & 1
\end{array}
$$

. . . .

where the sides of the triangle are made up of ones and each of the other numbers is the sum of the two numbers appearing above it (to the left and right), as indicated by the dashed lines. Show that the sum of the numbers in the $(n + 1)$st row of Pascal's triangle equals 2^n.

Hint. Show that the sum of the numbers in the $(n + 1)$st row is twice as large as the sum of those in the nth row.

55. What is the largest number of bishops that can be put on a chessboard without any two bishops threatening each other? Prove that this number of bishops can be put on in N different ways, where N is a perfect square.

Hint. To understand the problem, recall that in chess a bishop moves along a diagonal. For example, a bishop on the square d3 can go to any of the squares shown in Figure 5 (b1, c2, etc.) in a single move. Thus two bishops threaten each other if they occupy the same diagonal.

Fig. 5

Warning. The problem is sometimes "solved" by putting 14 bishops on a chessboard as shown in Figure 6. It is then claimed that if more than 14 bishops are put on the board, then two will necessarily threaten each other. This seems plausible, but not completely convincing. In any event, this approach does not lead to a solution of the second (and basic) part of the problem.

Fig. 6

56. A mother has two apples and three pears. Every day for five consecutive days she gives out one piece of fruit. How many ways can this be done?

Answer. $C_2^5 = 10$.

57. Answer the same question for k apples and n pears (given out one at a time over the course of $n + k$ days).

Answer. C_k^{k+n}. This answer will do for now, if you are unable to write an explicit formula for C_k^{k+n} in terms of n and k. Later on, we shall find such a formula (see Probs. 60, 66 and 67).

58. Answer the same question for 2 apples, 3 pears and 4 oranges (given out one at a time over the course of 9 days).

59. In how many ways can 8 rooks be put on a chessboard without any two rooks threatening each other?

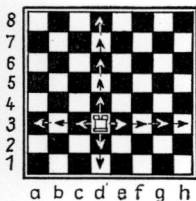

Fig. 7

Hint. A rook can move horizontally or vertically. For example, the rook on square d3 in Figure 7 threatens all squares in file d and row 3. Two rooks cannot be put in the same file (they would threaten each other), i.e., only one rook can be put in each file. But there are 8 files and 8 rooks. Hence there is exactly one rook in any given file. On how many different squares can this rook be placed?

60. How many distinct ways are there of choosing 2 objects from among n objects?

Hint. This quantity (also equal to the number of ways of turning on 2 bulbs out of n) has previously been denoted by C_2^n. Therefore the problem is to calculate C_2^n (for example, it has already been shown in Prob. 56 that $C_2^5 = 10$).

Answer. $C_2^n = \dfrac{n(n-1)}{2}$.

61. In how many ways can a class be seated if there are 26 students and 28 seats?

62. A father has five apples, all of different sizes, which he gives to his eight sons in such a way that each son either gets a whole apple or nothing. In how many ways can this be done?

Answer. $C_5^8 \cdot 5! = \dfrac{8!}{3!}$.

63. Let C_k^n denote the number of ways of lighting n bulbs such that k bulbs are on. Prove that C_k^n is the $(k+1)$st number in the $(n+1)$st row of Pascal's triangle (defined in Prob. 53).

64. In how many ways can four rooks be placed on a chessboard without any two threatening each other?

65. There are 3 officers and 40 men in a company. In how many ways can a patrol consisting of 1 officer and 3 men be chosen?

66. In how many ways can 3 objects be chosen from n objects (without regard for order)?

Answer. $C_3^n = \dfrac{n(n-1)(n-2)}{1 \cdot 2 \cdot 3}$.

67. In how many ways can k objects be chosen from n objects (without regard for order)?

Hint. Use the method of the preceding problem or mathematical induction.

Answer. $C_k^n = \dfrac{n(n-1)\cdots(n-k+1)}{1 \cdot 2 \cdots k}$.

68. In how many ways can six different consecutive notes be played on a piano? How many six-note chords are there?

Hint. A piano has 88 keys. In a chord all notes are played simultaneously.

69. How many terms are obtained after multiplying out the expression

$$(1 + a)(1 + b)(1 + c)(1 + d)(1 + e)(1 + f)(1 + g)?$$

Answer. 2^7.

70. How many of the terms obtained in the preceding problem are products of three letters?

Answer. $C_3^7 = \dfrac{7 \cdot 6 \cdot 5}{1 \cdot 2 \cdot 3} = 35.$

71. How many terms are obtained if we multiply on the expression

$$(1 + x + y)^{20}$$

without combining similar terms?

Answer. 3^{20}.

72. Suppose the expression

$$(1 + x^5 + x^7)^{20}$$

is multiplied out and similar terms are combined. What are the coefficients of x^{17} and x^{18}?

73. Suppose the expression

$$(1 + x)^{56}$$

is multiplied out and similar terms are combined. What are the coefficients of x^8 and x^{48}?

Answer. The coefficient of x^8 is C_8^{56}, and that of x^{48} is $C_{48}^{56} = C_8^{56}$.

74. Prove that

$$(1 + x)^n = C_0^n + C_1^n x + C_2^n x^2 + \cdots + C_n^n x^n,$$

a special case of the *binomial theorem* (see Prob. 77). Correspondingly, the numbers C_k^n ($k = 0, 1, \ldots, n$) are called the *binomial coefficients*.

75. Let $a_0, a_1, a_2, \ldots, a_n$ be the numbers in the $(n + 1)$st row of Pascal's triangle. Then it follows from Problems 63 and 74 that

$$(1 + x)^n = a_0 + a_1 x + a_2 x^2 + \cdots + a_{n-1} x^{n-1} + a_n x^n.$$

Give another proof of this fact, starting directly from the definition of Pascal's triangle.

76. Suppose k zeros are inserted between every pair of consecutive digits of the number 14641. Prove that the resulting number is a perfect square.

77. Suppose the expression

$$(a + b)^n$$

is multiplied out and similar terms are combined. Find the term involving a^k. Prove the *binomial theorem*

$$(a + b)^n = C_0^n a^n + C_1^n a b^{n-1} + C_2^n a^2 b^{n-2}$$

$$+ \cdots + C_k^n a^k b^{n-k} + \cdots + C_n^n b^n.$$

Hint. $(a + b)^n = a^n \left(1 + \dfrac{b}{a} \right)^n.$

78. In the expression $(x + y + z)^n$ find the term involving $x^k y^l$.

Hint. First find all terms involving x^k.

79. Calculate the sums

$$C_0^n + C_2^n + C_4^n + \cdots$$

and
$$C_1^n + C_3^n + C_5^n + \cdots.$$

Hint. Recall from Problem 45 that

$$2^n = C_0^n + C_1^n + C_2^n + \cdots + C_n^n,$$

but this time prove the formula by setting $x = 1$ in the binomial theorem
$$(1 + x)^n = C_0^n + C_1^n x + C_2^n x^2 + \cdots + C_n^n x^n$$

(see Prob. 74).

80. Find the sum of the coefficients of the polynomial obtained when the expression

$$(1 + x - 3x^2)^{1965}$$

is multiplied out and similar terms are combined.

Answer. -1.

81. How many diagonals can be drawn in a convex polygon with n sides?

Answer. $C_2^n - n$.

82. How many distinct even five-digit numbers can be formed from the digits 0, 1, 3, 4 and 5 if no repetitions are allowed?

Answer. 42.

83. How many distinct ways are there of putting a penny, a nickel, a dime, a quarter, a half-dollar and a silver dollar in two pockets?

84. How many distinct divisors does the number $2 \cdot 3 \cdot 5 \cdot 7 \cdot 11$ have?

85. Prove that

$$C_1^{n+1} + C_2^{n+2} + \cdots + C_k^{n+k} = C_k^{n+k+1} - 1.$$

86. How many ways are there of distributing $3n$ objects among three people in such a way that each person receives n objects?

87. A room contains several people, of whom 6 know English, 7 know French and 6 know German. Two know both English and French, 4 know both English and German, 3 know both French and German and one person knows all three languages. How many people are there in the room? How many know only English?

Answer. There are 11 people in the room, one of whom knows only English.

88. A library has a certain number of subscribers, each of whom has borrowed at least one book. Suppose that for every set of k books, it is known how many people have borrowed them at one time or another (here k ranges from 1 to n, the total number of books in the library). How many subscribers are there?

89. How many six-digit telephone numbers are there containing the combination 12?

Answer. 49401.

Detailed Solutions

1. The problem is solved by using mathematical induction:

a) The formula

$$1^3 + 2^3 + 3^3 + \cdots + n^3 = \frac{n^2 (n + 1)^2}{4} \qquad (1)$$

obviously holds for $n = 1$.

b) Suppose (1) holds for $n = k$, where k is an arbitrary positive integer, so that

$$1^3 + 2^3 + 3^3 + \cdots + k^3 = \frac{k^2 (k + 1)^2}{4}. \qquad (2)$$

Then we want to show that (1) also holds for $n = k + 1$, i.e., that

$$1^3 + 2^3 + 3^3 + \cdots + k^3 + (k + 1)^3 = \frac{(k + 1)^2 (k + 2)^2}{4}.$$

But

$$1^3 + 2^3 + 3^3 + \cdots + k^3 + (k + 1)^3$$

$$= \frac{k^2(k + 1)^2}{4} + (k + 1)^3 = (k + 1)^2 \left(\frac{k^2}{4} + k + 1 \right)$$

$$= \frac{(k + 1)^2}{4} (k^2 + 4k + 4) = \frac{(k + 1)^2 (k + 2)^2}{4}.$$

Therefore if (1) holds for $n = k$, it also holds for $n = k + 1$, and hence, by mathematical induction, (1) holds for all n.

2. First solution

a) The formula

$$\frac{1}{1 \cdot 2} + \frac{1}{2 \cdot 3} + \cdots + \frac{1}{n (n + 1)} = \frac{n}{n + 1} \qquad (1)$$

obviously holds for $n = 1$.

3 Silverman V

25

b) Suppose (1) holds for $n = k$, where k is an arbitrary positive integer, i.e., let

$$\frac{1}{1 \cdot 2} + \frac{1}{2 \cdot 3} + \cdots + \frac{1}{k(k+1)} = \frac{k}{k+1}.$$

Then (1) also holds for $n = k + 1$. In fact,

$$\frac{1}{1 \cdot 2} + \frac{1}{2 \cdot 3} + \cdots + \frac{1}{k(k+1)} + \frac{1}{(k+1)(k+2)}$$

$$= \frac{k}{k+1} + \frac{1}{(k+1)(k+2)} = \frac{k(k+2)+1}{(k+1)(k+2)}$$

$$= \frac{k^2 + 2k + 1}{(k+1)(k+2)} = \frac{(k+1)^2}{(k+1)(k+2)} = \frac{k+1}{k+2}.$$

Therefore, by mathematical induction, (1) holds for all n.

Second solution. By inspection,

$$\frac{1}{n(n+1)} = \frac{1}{n} - \frac{1}{n+1}$$

for any positive integer n. Therefore the left-hand side of (1) can be written as

$$\frac{1}{1 \cdot 2} + \frac{1}{2 \cdot 3} + \frac{1}{3 \cdot 4} + \cdots + \frac{1}{(n-1)n} + \frac{1}{n(n+1)}$$

$$= \left(\frac{1}{1} - \frac{1}{2}\right) + \left(\frac{1}{2} - \frac{1}{3}\right) + \left(\frac{1}{3} - \frac{1}{4}\right) + \cdots + \left(\frac{1}{n-1} - \frac{1}{n}\right)$$

$$+ \left(\frac{1}{n} - \frac{1}{n+1}\right)$$

$$= 1 - \frac{1}{2} + \frac{1}{2} - \frac{1}{3} + \frac{1}{3} - \frac{1}{4} + \cdots + \frac{1}{n-1} - \frac{1}{n} + \frac{1}{n} - \frac{1}{n+1}.$$

Since all the terms except the first and the last cancel, this sum reduces to

$$\frac{1}{1\cdot 2} + \frac{1}{2\cdot 3} + \cdots + \frac{1}{n\,(n+1)} = 1 - \frac{1}{n+1} = \frac{n}{n+1},$$

as in the first solution.

3. First solution

The formula

$$\frac{1}{1\cdot 4} + \frac{1}{4\cdot 7} + \cdots + \frac{1}{(3n-2)\,(3n+1)} = \frac{n}{3n+1} \qquad (1)$$

is easily verified by mathematical induction:

a) The formula holds for $n = 1$.

b) Suppose (1) holds for an arbitrary positive integer $n = k$, i.e., let

$$\frac{1}{1\cdot 4} + \frac{1}{4\cdot 7} + \cdots + \frac{1}{(3k-2)\,(3k+1)} = \frac{k}{3k+1}.$$

Then (1) also holds for $n = k + 1$. In fact

$$\frac{1}{1\cdot 4} + \frac{1}{4\cdot 7} + \cdots + \frac{1}{(3k-2)\,(3k+1)} + \frac{1}{(3k+1)\,(3k+4)}$$

$$= \frac{k}{3k+1} + \frac{1}{(3k+1)\,(3k+4)} = \frac{k\,(3k+4)+1}{(3k+1)\,(3k+4)}$$

$$= \frac{3k^2+4k+1}{(3k+1)\,(3k+4)} = \frac{(3k+1)\,(k+1)}{(3k+1)\,(3k+4)} = \frac{k+1}{3k+4}.$$

Therefore (1) holds for all n.

Second solution

Starting from the identity

$$\frac{1}{(3n-2)\,(3n+1)} = \frac{1}{3\,(3n-2)} - \frac{1}{3\,(3n+1)},$$

we write the left-hand side of (1) as

$$\frac{1}{1 \cdot 4} + \frac{1}{4 \cdot 7} + \frac{1}{7 \cdot 10} + \cdots + \frac{1}{(3n - 5)(3n - 2)}$$

$$+ \frac{1}{(3n - 2)(3n + 1)} = \left(\frac{1}{3 \cdot 1} - \frac{1}{3 \cdot 4} \right)$$

$$+ \left(\frac{1}{3 \cdot 4} - \frac{1}{3 \cdot 7} \right) + \left(\frac{1}{3 \cdot 7} - \frac{1}{3 \cdot 10} \right) + \cdots$$

$$+ \left(\frac{1}{3(3n - 5)} - \frac{1}{3(3n - 2)} \right) + \left(\frac{1}{3(3n - 2)} - \frac{1}{3(3n + 1)} \right).$$

Since all the terms except the first and the last cancel, this sum reduces to

$$\frac{1}{1 \cdot 4} + \frac{1}{4 \cdot 7} + \cdots + \frac{1}{(3n - 5)(3n - 2)}$$

$$= \frac{1}{3} - \frac{1}{3(3n + 1)} = \frac{n}{3n + 1},$$

as in the first solution.

4. One way of solving the problem involves the use of mathematical induction:

a) The formula

$$\frac{1}{a(a + 1)} + \frac{1}{(a + 1)(a + 2)} + \cdots + \frac{1}{(a + n - 1)(a + n)}$$

$$= \frac{n}{a(a + n)} \qquad (1)$$

obviously holds for $n = 1$.

b) Suppose (1) holds for an arbitrary positive integer $n = k$, so that

$$\frac{1}{a(a + 1)} + \frac{1}{(a + 1)(a + 2)} + \cdots + \frac{1}{(a + k - 1)(a + k)}$$

$$= \frac{k}{a(a + k)}.$$

Then (1) also holds for $n = k + 1$. In fact,

$$\frac{1}{a(a + 1)} + \frac{1}{(a + 1)(a + 2)} + \cdots + \frac{1}{(a + k - 1)(a + k)}$$

$$+ \frac{1}{(a + k)(a + k + 1)}$$

$$= \frac{k}{a(a + k)} + \frac{1}{(a + k)(a + k + 1)}$$

$$= \frac{k(a + k + 1) + a}{a(a + k)(a + k + 1)} = \frac{k^2 + ak + a + k}{a(a + k)(a + k + 1)}$$

$$= \frac{(a + k)(k + 1)}{a(a + k)(a + k + 1)} = \frac{k + 1}{a(a + k + 1)}.$$

Therefore (1) holds for all n.

5. We again use mathematical induction:

a) The formula

$$1 \cdot 1! + 2 \cdot 2! + \cdots + n \cdot n! = (n + 1)! - 1$$

holds for $n = 1$.

b) Suppose (1) holds for $n = k$, i.e.,

$$1 \cdot 1! + 2 \cdot 2! + \cdots + k \cdot k! = (k + 1)! - 1.$$

Then (1) also holds for $n = k + 1$. To see this, we merely note that
$$1 \cdot 1! + 2 \cdot 2! + \cdots + k \cdot k! + (k + 1)(k + 1)!$$

$$= (k + 1)! - 1 + (k + 1)(k + 1)! = (k + 2)(k + 1)! - 1.$$

But
$$(k + 2)(k + 1)! = (k + 2)!$$

by the definition of $n!$, and hence

$$1 \cdot 1! + 2 \cdot 2! + \cdots + (k + 1)(k + 1)! = (k + 2)! - 1.$$

It follows that (1) holds for all n.

6. First solution

a) The formula

$$\left(1 - \frac{1}{4}\right)\left(1 - \frac{1}{9}\right)\cdots\left(1 - \frac{1}{n^2}\right) = \frac{n+1}{2n} \tag{1}$$

obviously holds for $n = 2$.

b) Suppose (1) holds for any integer $k \geqslant 2$, i.e., let

$$\left(1 - \frac{1}{4}\right)\left(1 - \frac{1}{9}\right)\cdots\left(1 - \frac{1}{k^2}\right) = \frac{k+1}{2k} \, .$$

Then (1) also holds for $n = k + 1$. In fact

$$\left(1 - \frac{1}{4}\right)\left(1 - \frac{1}{9}\right)\cdots\left(1 - \frac{1}{k^2}\right)\left(1 - \frac{1}{(k+1)^2}\right)$$

$$= \frac{k+1}{2k}\left(1 - \frac{1}{(k+1)^2}\right) = \frac{(k+1)\,[(k+1)^2 - 1]}{2k\,(k+1)^2}$$

$$= \frac{(k+1)\,k\,(k+2)}{2k\,(k+1)^2} = \frac{k+2}{2\,(k+1)} \, .$$

Applying mathematical induction, we find that (1) holds for all $n \geqslant 2$.

Second solution

Starting from the identity

$$1 - \frac{1}{n^2} = \frac{(n-1)\,(n+1)}{n^2} \, ,$$

we write the left-hand side of (1) as

$$\left(1 - \frac{1}{4}\right)\left(1 - \frac{1}{9}\right)\left(1 - \frac{1}{16}\right)\cdots\left(1 - \frac{1}{(n-1)^2}\right)\left(1 - \frac{1}{n^2}\right)$$

$$= \frac{(2-1)\,(2+1)}{2^2}\,\frac{(3-1)\,(3+1)}{3^2}\,\frac{(4-1)\,(4+1)}{4^2}$$

$$\cdots \frac{(n-2)\,n}{(n-1)^2}\,\frac{(n-1)\,(n+1)}{n^2} \, .$$

Since all the factors except $2 - 1 = 1$ and $n + 1$ in the numerator and 2 and n in the denominator cancel, this product reduces to

$$\left(1 - \frac{1}{4}\right)\left(1 - \frac{1}{9}\right) \cdots \left(1 - \frac{1}{n^2}\right) = \frac{n + 1}{2n},$$

as in the first solution.

7. Use mathematical induction:

a) The formula

$$1 - 2^2 + 3^2 - \cdots + (-1)^{n-1} n^2 = (-1)^{n-1} \frac{n(n+1)}{2} \quad (1)$$

clearly holds for $n = 1$.

b) Suppose (1) holds for any positive integer k, so that

$$1 - 2^2 + 3^2 - \cdots + (-1)^{k-1} k^2 = (-1)^{k-1} \frac{k(k+1)}{2}.$$

Then (1) also holds for $n = k + 1$, since

$$1 - 2^2 + 3^2 - \cdots + (-1)^{k-1} k^2 + (-1)^k (k+1)^2$$

$$= (-1)^{k-1} \frac{k(k+1)}{2} + (-1)^k (k+1)^2$$

$$= (-1)^{k-1} (k+1) \left(\frac{k}{2} - k - 1\right)$$

$$= \frac{(-1)^{k-1}(k+1)}{2}(-k-2) = (-1)^k \frac{(k+1)(k+2)}{2}.$$

Therefore (1) holds for all n.

8. Assuming that n lines have already been drawn, we draw the $(n + 1)$st line. This line intersects the other n lines in just n points, since by hypothesis it intersects each of the other lines but does not intersect any two of them in the same point. It is clear that these n points of intersection divide the $(n + 1)$st line into $n + 1$ segments (two of which are infinite). Each of these segments divides an "old" part of the plane into two "new"

parts. Therefore the $n + 1$ segments together create $2n + 2$ new parts out of $n + 1$ old parts of the plane, i.e., drawing the $(n + 1)$st line increases the number of parts of the plane by $n + 1$.

It is now an easy matter to calculate the number of parts into which n lines divide the plane. One line divides the plane into 2 parts, drawing a second line increases the number of parts by 2, drawing a third line increases the number of parts by 3, and so on. Therefore the total number of parts equals

$$2 + 2 + 3 + 4 + \cdots + n = 1 + (1 + 2 + 3 + \cdots + n)$$

$$= 1 + \frac{n(n + 1)}{2}$$

(recall Example 1, p. 4). Therefore n lines divide the plane into

$$1 + \frac{n(n + 1)}{2} = \frac{n^2 + n + 2}{2}$$

parts.

9. The proof is by mathematical induction:

a) For $n = 8$ we have $8 = 3 + 5$.

b) Suppose k is a sum of threes and fives. Then this sum either contains a five (possibly several) or it does not. In the first case, replace a five by 2 threes. Then the new numbers are again all threes and fives and add up to $k + 1$. In the second case, there are at least 3 threes, and we can replace 3 threes by 2 fives. The new numbers are again all threes and fives and again add up to $k + 1$. In either case, if k can be written as a sum of threes and fives exclusively, so can $k + 1$. It follows by mathematical induction that any integer greater than 7 equals a sum of threes and fives.

10. The proof is again by induction:

a) If $n = 1$, so that the plane is divided by just one line, then there is an obvious regular coloring. In fact, we need only color one half-plane white and the other half-plane black.

b) Suppose there is a regular coloring of the plane divided by n lines, and draw the $(n + 1)$st line. This line divides the plane into two half-planes. In one of these half-planes, change the color of every part (i.e., change white to black and black to white), and in the other half-plane leave the color of every part the same. Then the new coloring is regular. To see this, consider any two adjacent parts of the plane divided by $n + 1$ lines. The line separating the two parts is either one of the old lines or else it is the new $(n + 1)$st line. In the first case, the two parts had different colors before drawing the new line and changing colors in one of the resulting half-planes. Therefore they either change color or they do not, but in any event they still have different colors. In the second case, the pieces had the same color before drawing the new line, but after drawing the new line one of the pieces has its color changed with the result that the two pieces now have different colors.

Thus if the plane divided by n lines can be regularly colored, so can the plane divided by $n + 1$ lines. It follows by mathematical induction that given any positive integer n, there is a regular coloring of the plane divided by n lines.

11. We must verify the assertion that

$$n^3 + (n + 1)^3 + (n + 2)^3$$

is divisible by 9 for every positive integer n. As usual, the proof is in two steps:

a) The assertion is true for $n = 1$, since $1^3 + 2^3 + 3^3 = 36$ is divisible by 9.

b) Suppose the assertion is true for a positive integer $n = k$, so that

$$k^3 + (k + 1)^3 + (k + 2)^3$$

is divisible by 9. Then the assertion is true for $n = k + 1$. In fact

$$(k + 1)^3 + (k + 2)^3 + (k + 3)^3$$

$$= (k + 1)^3 + (k + 2)^3 + (k^3 + 3 \cdot 3k^2 + 3 \cdot 3^2 k + 3^3)$$

$$= [k^3 + (k + 1)^3 + (k + 2)^3] + [9 (k^2 + 3k + 3)],$$

where both terms in brackets are divisible by 9, the first by hypothesis, the second since it is obviously a multiple of 9. Therefore the sum of the terms is also divisible by 9, i.e., the assertion is true for $n = k + 1$. It follows by mathematical induction that the assertion is true for all n.

12. This time the induction goes as follows:

a) For $n = 0$ the number

$$11^{n+2} + 12^{2n+1} = 11^2 + 12 = 121 + 12 = 133$$

is trivially divisible by 133.

b) If the number $11^{n+2} + 12^{2n+1}$ is divisible by 133 for some $n = k$, then it is divisible by 133 for $n = k + 1$. In fact

$$11^{(k+1)+2} + 12^{2(k+1)+1} = 11^{k+3} + 12^{2k+3}$$

$$= 11 \cdot 11^{k+2} + 144 \cdot 12^{2k+1}$$

$$= 11 \cdot 11^{k+2} + 11 \cdot 12^{2k+1} + 133 \cdot 12^{2k+1}$$

$$= 11 (11^{k+2} + 12^{2k+1}) + 133 \cdot 12^{2k+1},$$

where the last expression on the right is obviously divisible by 133, being the sum of two terms divisible by 133 (the first by hypothesis, the second by inspection). Therefore $11^{n+2} + 12^{2n+1}$ is divisible by 133 for $n = k + 1$. It follows by mathematical induction that $11^{n+2} + 12^{2n+1}$ is divisible by 133 for any positive integer n.

13. Use mathematical induction, starting from $n = 2$:

a) The inequality

$$(1 + a)^n > 1 + na \tag{1}$$

obviously holds for $n = 2$. In fact

$$(1 + a)^2 = 1 + 2a + a^2 > 1 + 2a,$$

since $a^2 > 0$ if $a \neq 0$.

b) Suppose (1) holds for some positive integer $k \geqslant 2$, i.e., let

$$(1 + a)^k > 1 + ka. \tag{2}$$

Then (1) also holds for $n = k + 1$. To see this, note that $1 + a > 0$ by hypothesis and multiply both sides of (2) by $1 + a$, obtaining

$$(1 + a)^k (1 + a) > (1 + ka) (1 + a)$$

or

$$(1 + a)^{k+1} > [1 + (k + 1) a] + ka.$$

But $ka^2 > 0$ if $a \neq 0$, and hence

$$(1 + a)^{k+1} > 1 + (k + 1) a.$$

It now follows by mathematical induction that (1) holds for all $n \geqslant 2$ if $a > -1$, $a \neq 0$.

14. The proof (again by induction) goes as follows:

a) The inequality

$$1 + \frac{1}{\sqrt{2}} + \frac{1}{\sqrt{3}} + \cdots + \frac{1}{\sqrt{n}} > \sqrt{n} \qquad (1)$$

holds for $n = 2$, since then it becomes

$$1 + \frac{1}{\sqrt{2}} > \sqrt{2}$$

which is equivalent to the obvious inequality

$$\sqrt{2} + 1 > 2.$$

b) Suppose (1) holds for $n = k$, so that

$$1 + \frac{1}{\sqrt{2}} + \frac{1}{\sqrt{3}} + \cdots + \frac{1}{\sqrt{k}} > \sqrt{k}. \qquad (2)$$

Then (1) also holds for $n = k + 1$. This is an immediate consequence of the inequality

$$\frac{1}{\sqrt{k + 1}} > \sqrt{k + 1} - \sqrt{k}, \qquad (3)$$

as we see by adding (2) and (3). To verify (3), we merely multiply both sides by $\sqrt{k + 1} + \sqrt{k}$, obtaining the inequality

$$1 + \sqrt{\frac{k}{k + 1}} > 1$$

which is obviously true. The validity of (1) for all $n \geqslant 2$ now follows by mathematical induction.

15. The result follows by mathematical induction, once we note that

a) $u_n = 2^{n-1} + 1$ for $n = 1$ and $n = 2$.

b) If $u_n = 2^{n-1} + 1$ holds for $n \leqslant k$ $(k \geqslant 2)$, then it also holds for $n = k + 1$. In fact

$$u_{k+1} = 3u_k - 2u_{k-1} = 3(2^{k-1} + 1) - 2(2^{k-2} + 1)$$
$$= 3 \cdot 2^{k-1} + 3 - 2^{k-1} - 2$$
$$= 2 \cdot 2^{k-1} + 1 = 2^k + 1.$$

16. Again the result follows by mathematical induction, after noting that

a) $u_n = (2n - 1)^2$ for $n = 1$.

b) If $u_n = (2n - 1)^2$ holds for $n = k$, then it also holds for $n = k + 1$. In fact

$$u_{k+1} = u_k + 8k = (2k - 1)^2 + 8k = 4k^2 - 4k + 1 + 8k$$
$$= 4k^2 + 4k + 1 = (2k + 1)^2 = [2(k + 1) - 1]^2.$$

17. $\Delta u_n = u_{n+1} - u_n = (n + 1)^2 - n^2 = 2n + 1$.

18. Since obviously

$$u_1 + (u_2 - u_1) + (u_3 - u_2) + \cdots + (u_n - u_{n-1}) = u_n$$

(all the terms cancel except the last), the formula

$$u_1 + \Delta u_1 + \Delta u_2 + \cdots + \Delta u_{n-1} = u_n \qquad (1)$$

is an obvious consequence of the definition of the first differences. Using (1) and the analogous formula for v_n, we have

$$u_n - v_n = (u_1 + \Delta u_1 + \Delta u_2 + \cdots + \Delta u_{n-1})$$
$$- (v_1 + \Delta v_1 + \Delta v_2 + \cdots + \Delta v_{n-1})$$
$$= u_1 - v_1 + \Delta u_1 - \Delta v_1 + \Delta u_2 - \Delta v_2 + \cdots$$
$$+ \Delta u_{n-1} - \Delta v_{n-1}.$$

By hypothesis

$$\Delta u_1 = \Delta v_1, \Delta u_2 = \Delta v_2, \ldots, \Delta u_{n-1} = \Delta v_{n-1},$$

and hence

$$u_n - v_n = u_1 - v_1.$$

It follows that if $u_1 - v_1 \neq 0$, then $u_n - v_n \neq 0$, i.e., u_n differs from v_n for every n. On the other hand, if $u_1 = v_1$, then $u_n = v_n$ for all n.

19. It need only be observed that

$$\Delta w_n = (w_{n+1} - w_n) = (u_{n+1} + v_{n+1}) - (u_n + v_n)$$

$$= (u_{n+1} - u_n) + (v_{n+1} - v_n) = \Delta u_n + \Delta v_n.$$

20. Noting that

$$\Delta u_n = (n + 1)^k - n^k,$$

we first show that

$$(n + 1)^k = n^k + kn^{n-1} + \ldots, \tag{1}$$

where the dots denote terms in n of degree less than $k - 1$. This is done by mathematical induction in k:

a) Formula (1) holds for $k = 1$.

b) Suppose (1) holds for $k = k_0$, i.e., let

$$(n + 1)^{k_0} = n^{k_0} + k_0 n^{k_0-1} + \ldots,$$

where the dots denote terms in n of degree less than $k_0 - 1$. Then (1) also holds for $k = k_0 + 1$. In fact

$$(n + 1)^{k_0+1} = (n + 1)^{k_0} (n + 1)$$

$$= (n^{k_0} + k_0 n^{k_0-1} + \cdots)(n + 1)$$

$$= n^{k_0+1} + k_0 n^{k_0} + \cdots + n^{k_0} + k_0 n^{k_0-1} + \cdots$$

$$= n^{k_0+1} + (k_0 + 1) n^{k_0} + \ldots,$$

where the last set of dots denotes terms in n of degree less than k_0. Therefore (1) holds for all k.

It now follows from (1) that

$$\Delta u_n = (n + 1)^k - n^k = n^k + kn^{k-1} + \cdots - n^k$$

$$= kn^{k-1} + \ldots,$$

where the last set of dots denotes terms in n of degree less than $k - 1$. In other words, Δu_n is a polynomial in n of degree $k - 1$, with leading coefficient equal to k.

21. Let

$$u_n = a_0 n^k + a_1 n^{k-1} + \cdots + a_{k-1} n + a_k,$$

and regard the sequence $\{u_n\}$ as the sum of $k + 1$ sequences

$$\{a_0 n^k\}, \{a_1 n^{k-1}\}, \ldots, \{a_{k-1}n\}, \{a_k\},$$

respectively. Then, according to Problem 19, $\{\Delta u_n\}$ is the sum of the first differences of these $k + 1$ sequences. Moreover, it follows from Problem 20 that the general term of the first sequence is a polynomial of degree $k - 1$ in n with leading coefficient $a_0 k$, while the general term of the remaining sequences are all polynomials of degree less than $k - 1$. Therefore Δu_n is a polynomial of degree $k - 1$ in n, with leading coefficient $a_0 k$.

22. If $u_n = n^k$, then the kth difference $\Delta^k u_n = k!$ This can be seen by using induction in k:

a) If $k = 1$, then $u_n = n$ and hence

$$\Delta u_n = u_{n+1} - u_n = (n + 1) - n = 1 = 1!$$

b) Suppose $\Delta^k u_n = k!$ for all $k \leqslant k_0$. Then $\Delta^k u_n = k!$ for $k = k_0 + 1$. To see this, consider the sequence $\{u_n\} = \{n^{k_0+1}\}$. Then, as we know from Problem 20, Δu_n is a polynomial in n of degree k_0 with leading coefficient $k_0 + 1$, say

$$\Delta u_n = (k_0 + 1) n^{k_0} + a_1 n^{k_0-1} + a_2 n^{k_0-2} + \cdots + a_{k_0-1} n + a_{k_0}.$$

But the sequence $\{\Delta u_n\}$ is the sum of the $k_0 + 1$ sequences

$$\{(k_0 + 1) n^{k_0}\}, \{a_1 n^{k_0-1}\}, \{a_2 n^{k_0-2}\}, \ldots, \{a_{k_0-1}n\}, \{a_{k_0}\},$$

respectively, and hence

$$\Delta^{k_0+1} u_n = \Delta^{k_0} \Delta u_n = \Delta^{k_0} (k_0 + 1) n^{k_0} + \Delta^{k_0} a_1 n^{k_0-1} + \Delta^{k_0} a_2 n^{k_0-1}$$

$$+ \cdots + \Delta^{k_0} a_{k_0-1} n + \Delta^{k_0} a_{k_0}$$

$$= (k_0 + 1) \Delta^{k_0} n^{k_0} + a_1 \Delta \Delta^{k_0-1} n^{k_0-1} + a_2 \Delta^2 \Delta^{k_0-2} n^{k_0-2}$$

$$+ \cdots + a_{k_0-1} \Delta^{k_0-1} \Delta n + \Delta^{k_0} a_{k_0} \qquad (1)$$

(justify this equation!). Moreover, by hypothesis,

$$\Delta^k n^k = k! \quad \text{for all} \quad k \leqslant k_0,$$

and hence (1) reduces

$$\Delta^{k_0+1} u_n = (k_0 + 1) k_0! + a_1 \Delta (k_0 - 1)! + a_2 \Delta^2 (k_0 - 2)!$$

$$+ \cdots + a_{k_0-1} \Delta^{k_0-1} 1 + \Delta^{k_0} a_{k_0}$$

$$= (k_0 + 1) k_0! = (k_0 + 1)!$$

since all the other terms vanish (if the terms of a sequence all equal the same constant, then its first differences, second differences, third differences, etc. all vanish). Therefore $\Delta^k u_n = k!$ for $k = k_0 + 1$, and the proof by induction is complete.

23. The proof is by induction:

a) Let $k = 0$ so that $v_n = 1$ for all n. Then the assertion is clearly true, the required sequence being just the sequence with general term $u_n = n$. Let this sequence be denoted by $\{u_n^{(0)}\}$.

b) Suppose the assertion is true for all $k \leqslant k_0$, so that there exist sequences $\{u_n^{(0)}\}, \{u_n^{(1)}\}, \ldots, \{u_n^{(k_0)}\}$ such that

$$\Delta u_n^{(0)} = 1, \Delta u_n^{(1)} = n, \ldots, \Delta u_n^{(k_0)} = n^{k_0}.$$

Then there exists a sequence $\{u_n^{(k_0+1)}\}$ such that

$$\Delta u_n^{(k_0+1)} = n^{k_0+1}. \qquad (1)$$

To find $u_n^{(k_0+1)}$, we start from the sequence with general term n^{k_0+2}. According to Problem 21, Δn^{k_0+2} is a polynomial of degree $k_0 + 1$ in n with leading coefficient $k_0 + 2$. Therefore

$$\Delta \left(\frac{n^{k_0+2}}{k_0 + 2} \right)$$

is a polynomial of degree $k_0 + 1$ in n with leading coefficient 1. Let this polynomial be

$$n^{k_0+1} + a_0 n^{k_0} + a_1 n^{k_0-1} + \cdots + a_{k_0}.$$

Then for $\{u^{(k_0+1)}\}$ we choose the sequence with general term

$$u_n^{k_0+1} = \frac{n^{k_0+2}}{k_0 + 2} - a_0 u_n^{(k_0)} - a_1 u_n^{(k_0-1)} - \cdots - a_{k_0} u_n^{(0)}.$$

To verify that this sequence has the required properties, we use Problem 19, obtaining

$$\Delta u_n^{(k_0+1)} = \Delta \left(\frac{n^{k_0+2}}{k_0 + 2} \right) - \Delta \left(a_0 u_n^{(k_0)} \right) - \Delta \left(a_1 u_n^{(k_0-1)} \right)$$

$$- \cdots - \Delta \left(a_{k_0} u_n^{(0)} \right)$$

$$= n^{k_0+1} + a_0 n^{k_0} + a_1 n^{k_0-1} + \cdots + a_{k_0}$$

$$- a_0 n^{k_0} - a_1 n^{k_0-1} - \cdots - a_{k_0} = n^{k_0+1},$$

which agrees with (1). The induction is now complete, i.e., for every k there exists a sequence $\{u^{(k)}\}$ such that $\Delta u_n^{(k)} = n^k$. Moreover, it is clear from the above construction that $u_n^{(k)}$ is a polynomial of degree $k + 1$ in n with leading coefficient $1/(k + 1)$.

24. Since $\Delta^4 u_n = 0$ by hypothesis, we have $\Delta^3 u_n = c$ (for all n) where c is a constant. Therefore $\Delta^2 u_n$ is a polynomial of degree 1 in n. Let this polynomial be $a_0 n + a_1$. Then, by the solution of Problem 18,

$$\Delta u_n = \Delta u_1 + \Delta^2 u_1 + \Delta^2 u_2 + \cdots + \Delta^2 u_{n-1}$$

$$= \Delta u_1 + a_1 (n - 1) + a_0 [1 + 2 + \cdots + (n - 1)].$$

But

$$1 + 2 + \cdots + (n - 1) = \frac{n(n - 1)}{2}$$

(see Example 1, p. 4), and hence

$$\Delta u_n = \Delta u_1 + a_1(n - 1) + a_0 \frac{n(n - 1)}{2},$$

i.e., Δu_n is a polynomial of degree 2 in n. Let this polynomial be $b_0 n^2 + b_1 n + b_2$. Then

$$u_n = u_1 + \Delta u_1 + \Delta u_2 + \cdots + \Delta u_{n-1}$$

$$= u_1 + b_2(n - 1) + b_1 \frac{n(n - 1)}{2} + b_0[1^2 + 2^2 + \cdots + (n - 1)^2].$$

If S_n denotes the sum in brackets, then $\Delta S_n = n^2$ and hence by Problem 23, S_n is a polynomial of degree 3 in n. Therefore u_n is also a polynomial of degree 3 in n.

25. Let

$$\Delta u_n = a_0 n^k + a_1 n^{k-1} + \cdots + a_{k-1} n + a_k,$$

and consider the sequence with general term

$$v_n = a_0 u_n^{(k)} + a_1 u_n^{(k-1)} + \cdots + a_{k-1} u_n^{(1)} + a_k u_n^{(0)},$$

where $u_n^{(k)}$, $u_n^{(k-1)}$, ..., $u_n^{(1)}$, $u_n^{(0)}$ are the same as in the solution of Problem 23. Then $\Delta v_n = \Delta u_n$ for all n. It follows from the solution of Problem 18a that $u_n - v_n = u_1 - v_1$ for all n. Therefore

$$u_n = (a_0 u_n^{(k)} + a_1 u_n^{(k-1)} + \cdots + a_{k-1} u_n^{(1)} + a_k u_n^{(0)}) + (u_1 - v_1).$$

But, according to Problem 23, $u_n^{(k)}$ is a polynomial of degree $k + 1$ in n, and hence so is u_n.

26. If

$$u_n = 1^2 + 2^2 + \cdots + n^2,$$

then

$$\Delta u_n = u_{n+1} - u_n = (n + 1)^2.$$

is a polynomial of degree 2 in n. Therefore, by Problem 25, u_n is a polynomial of degree 3 in n, say

$$u_n = a_0 n^3 + a_1 n^2 + a_2 n + a_3.$$

To find the coefficients a_0, a_1, a_2 and a_3, we replace n by 0, 1, 2 and 3, obtaining

$$u_0 = a_3 = 0,$$

$$u_1 = a_0 + a_1 + a_2 + a_3 = 1,$$

$$u_2 = 8a_0 + 4a_1 + 2a_2 + a_3 = 5,$$

$$u_3 = 27a_0 + 9a_1 + 3a_2 + a_3 = 14.$$

Hence

$$a_0 + a_1 + a_2 = 1,$$

$$8a_0 + 4a_1 + 2a_2 = 5,$$

$$27a_0 + 9a_1 + 3a_2 = 14,$$

and to solve this system we subtract three times the first equation from the third equation and twice the first equation from the second equation. This gives

$$24a_0 + 6a_1 = 11,$$

$$6a_0 + 2a_1 = 3,$$

which implies

$$6a_0 = 2, \quad a_0 = \tfrac{1}{3}, \quad a_1 = \tfrac{1}{2}, \quad a_2 = \tfrac{1}{6}.$$

It follows that

$$u_n = \tfrac{1}{3}n^3 + \tfrac{1}{2}n^2 + \tfrac{1}{6}n = \frac{n(2n^2 + 3n + 1)}{6}$$

$$= \frac{n(n+1)(2n+1)}{6},$$

i.e.,

$$1^2 + 2^2 + \cdots + n^2 = \frac{n(n+1)(2n+1)}{6}.$$

It should be noted that the answer can be verified by mathematical induction if the answer is known in advance or if it has been guessed somehow.

27. If
$$u_n = 1 \cdot 2 + 2 \cdot 3 + \cdots + n(n + 1),$$
then
$$\Delta u_n = (n + 1)(n + 2)$$

is a polynomial of degree 2 in n. It follows from Problem 25 that u_n is a polynomial of degree 3 in n, say

$$u_n = a_0 n^3 + a_1 n^2 + a_2 n + a_3.$$

To find the coefficients a_0, a_1, a_2 and a_3, we replace n by 0, 1, 2 and 3, obtaining

$$u_0 = a_3 = 0,$$

$$u_1 = a_0 + a_1 + a_2 + a_3 = 2,$$

$$u_2 = 8a_0 + 4a_1 + 2a_2 + a_3 = 8,$$

$$u_3 = 27a_0 + 9a_1 + 3a_2 + a_3 = 2.$$

Hence
$$a_0 + a_1 + a_2 = 2,$$

$$8a_0 + 4a_1 + 2a_2 = 8,$$

$$27a_0 + 9a_1 + 3a_2 = 20,$$

and to solve this system we subtract three times the first equation from the third equation and twice the first equation from second. This gives

$$24a_0 + 6a_1 = 14,$$

$$6a_0 + 2a_1 = 4,$$

which implies

$$6a_0 = 2, \quad a_0 = \tfrac{1}{3}, \quad a_1 = 1, \quad a_2 = \tfrac{2}{3}.$$

It follows that

$$u_n = \tfrac{1}{3} n^3 + n^2 + \tfrac{2}{3} n$$

$$= \frac{n (n^2 + 3n + 2)}{3} = \frac{n (n + 1) (n + 2)}{3},$$

i.e.,

$$1 \cdot 2 + 2 \cdot 3 + \cdots + n (n + 1) = \frac{n (n + 1) (n + 2)}{3}.$$

28. a) By the definition of an arithmetic progression,

$$u_n = u_{n-1} + d = u_{n-2} + 2d = \cdots = u_1 + (n - 1) d.$$

b) By the definition of a geometric progression,

$$u_n = q u_{n-1} = q^2 u_{n-2} = \cdots = q^{n-1} u_1.$$

29. a) According to Problem 28a,

$$S_n = u_1 + (u_1 + d) + \cdots + [u_1 + (n - 2) d]$$
$$+ [u_1 + (n - 1) d]. \tag{1}$$

Writing this sum in reverse order, we have

$$S_n = [u_1 + (n - 1) d] + [u_1 + (n - 2) d] + \cdots$$
$$+ (u_1 + d) + u_1. \tag{2}$$

We then add (1) and (2) term by term, obtaining

$$2S_n = \{u_1 + [u_1 + (n - 1) d\} + \{(u_1 + d) + [u_1 + (n - 2) d\}$$
$$+ \cdots + \{[u_1 + (n - 2) d] + (u_1 + d)\} + \{[u_1 + (n - 1) d] + u_1\}.$$

There are n terms in curly brackets, each equal to $2u_1 + (n - 1)d$.
Therefore

$$2S_n = n [2u_1 + (n - 1) d] = 2nu_1 + n (n - 1) d$$

or

$$S_n = nu_1 + \frac{n (n - 1)}{2} d.$$

b) According to Problem 28 b,

$$P_n = u_1 \cdot qu_1 \cdot q^2u_1 \cdots q^{n-2}u_1 \cdot q^{n-1}u_1$$
$$= u_1^n q^{1+2+\cdots+(n-2)+(n-1)}.$$

But by Example 1, p. 4,

$$1 + 2 + \cdots + (n-2) + (n-1) = \frac{n(n-1)}{2},$$

and hence

$$P_n = u_1^n q^{n(n-1)/2}.$$

30. a) Using the formula

$$S_n = nu_1 + \frac{n(n-1)}{2} d$$

found in Problem 29 a, we have

$$S_{15} = \frac{15 \cdot 14}{2} \frac{1}{3} = 35.$$

b) Using the formula

$$P_n = u_1^n q^{n(n-1)/2}$$

found in Problem 29 b, we have

$$P_{15} = \left(\sqrt[3]{10}\right)^{\frac{15 \cdot 14}{2}} = 10^{\frac{15 \cdot 14}{2} \frac{1}{3}} = 10^{35}.$$

31. a) In terms of the difference d, we have

$$u_1 = u_3 - 2d, u_2 = u_1 - d, u_4 = u_3 + d, u_5 = u_3 + 2d.$$

Therefore the sum of the first five terms equals

$$S_5 = (u_3 - 2d) + (u_3 - d) + u_3 + (u_3 + d) + (u_3 + 2d)$$
$$= 5u_3 = 0.$$

b) In terms of the ratio q, we have

$$u_1 = \frac{u_3}{q^2}, \quad u_2 = \frac{u_3}{q}, \quad u_4 = qu_3, \quad u_5 = q^2u_3.$$

Therefore the product of the first five terms equals

$$P_5 = \frac{u_3}{q^2} \frac{u_3}{q} u_3 q u_3 q^2 u_3 = u_3^5 = 4^5 = 1024.$$

32. It follows from

$$S_n = u_1 + u_2 + u_3 + \cdots + u_n$$

$$= u_1 + qu_1 + q^2 u_1 + \cdots + q^{n-1} u_1$$

that

$$qS_n - S_n = q\,(u_1 + qu_1 + q^2 u_1 + \cdots + q^{n-2} u_1 + q^{n-1} u_1)$$

$$- (u_1 + qu_1 + q^2 u_1 + q^3 u_1 + \cdots + q^{n-1} u_1)$$

$$= qu_1 + q^2 u_1 + q^3 u_1 + \cdots + q^{n-1} u_1 + q^n u_1$$

$$- u_1 - qu_1 - q^2 u_1 - q^3 u_1 - \cdots - q^{n-1} u_1$$

$$= q^n u_1 - u_1$$

(all but two terms cancel). Therefore

$$S_n\,(q - 1) = u_1\,(q^n - 1),$$

i.e.,

$$S_n = \frac{q^n - 1}{q - 1}\,u_1\,.$$

33. Ten minutes after the arrival of the messenger, 2 new people learn the news, ten minutes later 4 more people learn the news, and so on, with 2^k new people learning the news $10k$ minutes after the messenger's arrival. Therefore the number of people who know the news after $10k$ minutes equals the number who know the news when the messenger arrives (only the messenger himself!) plus those who learn the news 10 minutes later (2 people) plus those who learn the news 20 minutes later (4 more people), and so on, up to and including the number who learn the news $10k$ minutes later. This number is just

$$S_k = 1 + 2 + 4 + \cdots + 2^k,$$

i.e., the sum of the first $k + 1$ terms of a geometric progression with first term 1 and ratio 2, and according to the preceding problem,

$$S_k = \frac{2^{k+1} - 1}{2 - 1} \, 1 = 2^{k+1} - 1.$$

We must now find the smallest value of k for which S_k (the number of people who know the news after $10k$ minutes) exceeds 3,000,000 (the number of inhabitants of the city). This value of k is 21, since

$$2^{22} - 1 = 4,194,303.$$

Therefore the whole city knows the news after 210 minutes = 3 hours and 30 minutes.

34. Suppose it takes the bicyclist a_n seconds to complete the nth lap ($n = 1, 2, 3, 4, 5$), while it takes the horseback rider b_n seconds. Then the numbers a_1, a_2, a_3, a_4, a_5 form a geometric progression with ratio 1.1, while the numbers b_1, b_2, b_3, b_4, b_5 form an arithmetic progression with difference d. Moreover, by hypothesis, $a_1 = b_1$ and

$$a_1 + a_2 + a_3 + a_4 + a_5 = b_1 + b_2 + b_3 + b_4 + b_5. \quad (1)$$

Using the formulas found in Problems 29 and 32, we can write (1) as

$$\frac{(1.1)^5 - 1}{1.1 - 1} a_1 = 5a_1 + \frac{5 \cdot 4}{2} d,$$

which implies

$$a_1 \{ 10 \, [(1.1)^5 - 1] - 5 \} = 10d$$

or

$$\frac{d}{a_1} = \frac{10 \, [(1.1)^5 - 1] - 5}{10}. \quad (2)$$

We want to find the ratio

$$\frac{b_5}{a_5} = \frac{a_1 + 4d}{(1.1)^4 a_1} = \frac{1}{(1.1)^4} + \frac{4}{(1.1)^4} \frac{d}{a_1}. \tag{3}$$

Substitution of (2) into (3) gives

$$\frac{b_5}{a_5} = \frac{1}{(1.1)^4} + \frac{4}{(1.1)^4} \frac{10[(1.1)^5 - 1] - 5}{10}$$

$$= \frac{1}{(1.1)^4} \{1 + 4[(1.1)^5 - 1] - 2\}$$

$$= \frac{1}{(1.1)^4} [4(0.6105) - 1] = \frac{1.4420}{1.4641} \approx 0.985 < 1.$$

Therefore it takes the horseback rider 0.985 less time than the bicyclist to complete the last lap.

35. Clearly $S = 1 + 3 + 5 + \cdots + 997 + 999$ is the sum of the first 500 terms of an arithmetic progression with first term 1 and difference 2. Therefore

$$S = 500 \cdot 1 + \frac{500 \cdot 499}{2} 2 = 500 + 500 \cdot 499 = 500 \cdot 500$$

$$= 250{,}000.$$

36. Suppose that from $S^{(1)}$, the sum of all three-digit numbers not divisible by 2 or by 3, we subtract $S^{(2)}$, the sum of all three-digit numbers not divisible by 2, and $S^{(3)}$, the sum of all three-digit numbers not divisible by 3. In so doing, the numbers divisible by both 2 and 3, i.e., divisible by 6, have been subtracted twice. Therefore to obtain the answer, we must add $S^{(6)}$, the sum of all three-digit numbers divisible by 6, i.e.,

$$S = S^{(1)} - S^{(2)} - S^{(3)} + S^{(6)}.$$

Each of the quantities $S^{(1)}$, $S^{(2)}$, $S^{(3)}$ and $S^{(6)}$ is the sum of a certain number of terms of an arithmetic progression, as shown in the following table, where u_1 is the first term, n the number of terms, and the difference:

	u_1	n	d
$S^{(1)}$	100	900	1
$S^{(2)}$	100	450	2
$S^{(3)}$	102	300	3
$S^{(6)}$	102	150	6

Therefore

$$S^{(1)} = 900 \cdot 100 + \frac{900 \cdot 899}{2} \, 1 = 90{,}000 + 404{,}550 = 494{,}550,$$

$$S^{(2)} = 450 \cdot 100 + \frac{450 \cdot 449}{2} \, 2 = 45{,}000 + 202{,}050 = 247{,}050,$$

$$S^{(3)} = 300 \cdot 102 + \frac{300 \cdot 299}{2} \, 3 = 30{,}600 + 134{,}550 = 165{,}150,$$

$$S^{(6)} = 150 \cdot 102 + \frac{150 \cdot 149}{2} \, 6 = 15{,}300 + 67{,}050 = 82{,}350,$$

and hence finally

$$S = S^{(1)} - S^{(2)} - S^{(3)} + S^{(6)} = 164{,}700.$$

37. Obviously the nth term of any sequence equals the sum of the first n terms minus the sum of the first $n - 1$ terms, i.e.,

$$u_n = S_n - S_{n-1}.$$

In our case,

$$u_n = S_n - S_{n-1} = 3n^2 - 3(n - 1)^2 = 6n - 3.$$

Moreover

$$u_n - u_{n-1} = (6n - 3) - [6(n - 1) - 3] = 6$$

for all $n \geqslant 2$. Therefore the difference between consecutive terms of the sequence is constant and equals 6, i.e., the sequence is an arithmetic progression with difference 6. Its first term is clearly $u_1 = S_1 = 3$.

38. Suppose we have found a geometric progression with 27 as its mth term, 8 as its nth term and 12 as its pth term. Let u_1 denote the first term of the progression and q its ratio. Then

$$27 = q^{m-1}u_1, \quad 8 = q^{n-1}u_1, \quad 12 = q^{p-1}u_1,$$

and dividing the first equation by the second and third gives

$$\tfrac{27}{8} = q^{m-n}, \quad \tfrac{9}{4} = q^{m-p}.$$

Raising the first of these equations to the power $m - p$ and the second to the power $n - p$, we equate their right-hand sides, obtaining

$$(\tfrac{27}{8})^{m-p} = (\tfrac{9}{4})^{m-n} \qquad (1)$$

which implies

$$3^{3m-3p-2m+2n} = 2^{3m-3p-2m+2n}$$

or

$$3^{m-3p+2n} = 2^{m-3p+2n} \qquad (2)$$

Clearly (2) can hold only if $m - 3p + 2n = 0$ (why?).

Conversely, if m, n and p are distinct positive integers such that $m - 3p + 2n = 0$, then (2) and (1) hold. Let

$$q = \sqrt[m-n]{\tfrac{27}{8}} \, .$$

Then

$$\tfrac{27}{8} = q^{m-n}, \quad \tfrac{9}{4} = q^{m-p}. \qquad (3)$$

Let u_1 be such that $12 = q^{p-1}u_1$ (such a number always exists). Then it follows from (3) that

$$8 = q^{n-1}u_1, \quad 27 = q^{m-1}u_1.$$

This means that the numbers 27, 8 and 12 are the mth, nth and pth terms of a geometric progression with first term u_1 and ratio q. For example, if $u_1 = 8$, $q = \frac{3}{2}$, then $m = 4$, $n = 1$, $p = 2$.

39. Suppose we have found a geometric progression with 1 as its mth term, 2 as its nth term and 5 as its pth term. Let u_1 denote the first term of the progression and q its ratio. Then

$$1 = q^{m-1}u_1, \quad 2 = q^{n-1}u_1, \quad 5 = q^{p-1}u_1,$$

and dividing the second and third equations by the first gives

$$2 = q^{n-m}, \quad 5 = q^{p-m}.$$

Raising the first of these equations to the power $p - m$ and the second to the power $n - m$, we equate their right-hand sides, obtaining

$$2^{p-m} = 5^{n-m}.$$

Clearly this relation can hold only if $p - m = 0$ and $n - m = 0$ (why?). But this is impossible, since m, n and p must obviously be distinct. Hence there is no geometric progression with 1 as its mth term, 2 as its nth term and 5 as its pth term.

40. Let u_{12}, u_{13} and u_{15} denote the twelfth, thirteenth and fifteenth terms of the arithmetic progression. Then, by hypothesis, the numbers u_{12}^2, u_{13}^2, u_{15}^2 form a geometric progression, i.e.,

$$\frac{u_{13}^2}{u_{12}^2} = \frac{u_{15}^2}{u_{13}^2}.$$

Taking the square root of both sides of this equation, we find that either

$$\frac{u_{13}}{u_{12}} = \frac{u_{15}}{u_{13}} \tag{1}$$

or

$$\frac{u_{13}}{u_{12}} = -\frac{u_{15}}{u_{13}}. \tag{2}$$

First we consider (1). For any arithmetic progression with difference d, we have

$$u_{13} = u_{12} + d, \quad u_{15} = u_{13} + 2d = u_{12} + 3d,$$

and hence

$$\frac{u_{12} + d}{u_{12}} = \frac{u_{13} + 2d}{u_{13}},$$

i.e.,

$$1 + \frac{d}{u_{12}} = 1 + \frac{2d}{u_{13}}$$

or

$$\frac{d}{u_{12}} = \frac{2d}{u_{13}}. \tag{3}$$

Assuming that $d \neq 0$, we find from (3) that

$$\frac{u_{13}}{u_{12}} = 2, \quad \frac{u_{13}^2}{u_{12}^2} = 4,$$

i.e., the ratio of the geometric progression is 4.

Next we consider (2). In this case,

$$\frac{u_{12} + d}{u_{12}} = -\frac{u_{13} + 2d}{u_{13}},$$

i.e.,

$$1 + \frac{d}{u_{12}} = -1 - \frac{2d}{u_{13}}$$

or

$$2 + \frac{d}{u_{12}} + \frac{2d}{u_{13}} = 0.$$

It follows that

$$2u_{12}u_{13} + du_{13} + 2du_{12} = 0$$

or

$$2u_{12}^2 + 5du_{12} + d^2 = 0.$$

Therefore

$$\left(\frac{d}{u_{12}}\right)^2 + 5\,\frac{d}{u_{12}} + 2 = 0,$$

so that

$$\frac{d}{u_{12}} = \frac{-5 \pm \sqrt{17}}{2}.$$

There are now two possibilities:

a)

$$\frac{d}{u_{12}} = -\frac{5}{2} + \frac{\sqrt{17}}{2},$$

$$\frac{u_{13}}{u_{12}} = 1 + \frac{d}{u_{12}} = -\frac{3}{2} + \frac{\sqrt{17}}{2},$$

$$q = \left(\frac{u_{13}}{u_{12}}\right)^2 = \frac{13}{2} - \frac{3}{2}\sqrt{17}.$$

b)

$$\frac{d}{u_{12}} = -\frac{5}{2} - \frac{\sqrt{17}}{2},$$

$$\frac{u_{13}}{u_{12}} = 1 + \frac{d}{u_{12}} = -\frac{3}{2} - \frac{\sqrt{17}}{2},$$

$$q = \left(\frac{u_{13}}{u_{12}}\right)^2 = \frac{13}{2} + \frac{3}{2}\sqrt{17}.$$

In other words, (2) implies that q equals either $\frac{13}{2} + \frac{3}{2}\sqrt{17}$ or $\frac{13}{2} - \frac{3}{2}\sqrt{17}$.

There still remains the possibility that $d = 0$, but in this case it is obvious that $q = 1$.

41. If $\{u_n\}$ is a geometric progression such that

$$u_n = u_{n-1} + u_{n-2} \tag{1}$$

for all $n \geqslant 3$, then, by Problem 28b,

$$q^{n+1}u_1 = q^n u_1 + q^{n-1}u_1$$

for all $n \geqslant 1$. It follows that

$$q^2 = q + 1$$

and hence

$$q = \frac{1 \pm \sqrt{5}}{2}. \tag{2}$$

Conversely, it is easy to see that (2) implies (1).

42. Let the first progression have first term u_1 and ratio p, while the second progression has first term v_1 and ratio q. Moreover, let the given sequence be $\{w_n\}$. Then

$$w_1 = u_1 + v_1 = 0, \tag{1}$$

$$w_2 = pu_1 + qv_1 = 0, \tag{2}$$

by hypothesis. But (1) implies

$$v_1 = -u_1,$$

and then (2) in turn implies

$$pu_1 - qu_1 = (p - q)\, u_1 = 0$$

or $p = q$. It follows that

$$w_3 = p^2u_1 + q^2v_1 = p^2u_1 - q^2u_1 = q^2u_1 - q^2u_1 = 0,$$

i.e., the third term of the given sequence also equals zero.

43. By hypothesis,

$$u_1 + v_1 = 1, \quad u_2 + v_2 = 1$$

and

$$u_n + v_n = u_{n-1} + v_{n-1} + u_{n-2} + v_{n-2}. \tag{1}$$

Let p and q be the ratios of the progressions $\{u_n\}$ and $\{v_n\}$. Clearly $p \neq q$, since the sum of two geometric progressions with the same ratio is itself a geometric progression and the Fibonacci sequence is not a geometric progression. Equation (1) can be written in the form

$$p^{n-1}u_1 + q^{n-1}v_1 = p^{n-2}u_1 + q^{n-2}v_1 + p^{n-3}u_1 + q^{n-3}v_1$$

or
$$u_1 p^{n-3} (p^2 - p - 1) = -v_1 q^{n-3} (q^2 - q - 1). \qquad (2)$$

It follows that
$$p^2 - p - 1 = 0,$$
$$q^2 - q - 1 = 0.$$

In fact, suppose this is not true and let $q^2 - q - 1 \neq 0$, say. Then, dividing both sides of (2) by
$$v_1 p^{n-3} (q^2 - q - 1),$$

we obtain
$$\frac{u_1}{v_1} \frac{p^2 - p - 1}{q^2 - q - 1} = -\left(\frac{q}{p}\right)^{n-3} \qquad (3)$$

for all $n > 3$. Since the left-hand side of (3) does not depend on n, neither can the right-hand side. But this can be so only if $p = q$, and, as just noted, $p \neq q$. This contradiction shows that the assumption $q^2 - q - 1 \neq 0$ is false. Similarly, it can be shown that $p^2 - p - 1 = 0$. It follows that p and q are distinct roots of the equation $x^2 - x - 1 = 0$. This equation has roots

$$\frac{1 \pm \sqrt{5}}{2},$$

and hence
$$p = \frac{1 + \sqrt{5}}{2}, \quad q = \frac{1 - \sqrt{5}}{2},$$

say.

Finally, we find u_1 and v_1, using the pair of equations

$$u_1 + v_1 = 0,$$

$$\frac{1 + \sqrt{5}}{2} u_1 + \frac{1 - \sqrt{5}}{2} v_1 = 1,$$

with solution

$$u_1 = \frac{1 + \sqrt{5}}{2\sqrt{5}}, \quad v_1 = \frac{-1 + \sqrt{5}}{2\sqrt{5}}.$$

Therefore

$$u_n = p^{n-1}u_1 = \frac{1+\sqrt5}{2\sqrt5}\left(\frac{1+\sqrt5}{2}\right)^{n-1} = \frac{1}{\sqrt5}\left(\frac{1+\sqrt5}{2}\right)^n,$$

$$v_n = q^{n-1}v_1 = \frac{-1+\sqrt5}{2\sqrt5}\left(\frac{1-\sqrt5}{2}\right)^{n-1} = -\frac{1}{\sqrt5}\left(\frac{1-\sqrt5}{2}\right)^n.$$

and hence

$$w_n = u_n + v_n = \frac{1}{\sqrt5}\left[\left(\frac{1+\sqrt5}{2}\right)^n - \left(\frac{1-\sqrt5}{2}\right)^n\right].$$

44. First solution

The number of ways of lighting the bulbs such that one of the five bulbs is on equals the number of ways such that one bulb is off or equivalently such that four bulbs are on. In each case, there are 5 ways in all (the symbol $\sim$ means "corresponds to"):

```
○ ● ● ● ●   ~   ● ○ ○ ○ ○
● ○ ● ● ●   ~   ○ ● ○ ○ ○
● ● ○ ● ●   ~   ○ ○ ● ○ ○
● ● ● ○ ●   ~   ○ ○ ○ ● ○
● ● ● ● ○   ~   ○ ○ ○ ○ ●
```

Similarly, the number of ways of lighting the bulbs such that two bulbs are on equals the number of ways such that two bulbs are off or equivalently such that three bulbs are on. In each case, there are 10 ways in all:

```
○ ○ ● ● ●   ~   ● ● ○ ○ ○
○ ● ○ ● ●   ~   ● ○ ● ○ ○
○ ● ● ○ ●   ~   ● ○ ○ ● ○
○ ● ● ● ○   ~   ● ○ ○ ○ ●
● ○ ○ ● ●   ~   ○ ● ● ○ ○
● ○ ● ○ ●   ~   ○ ● ○ ● ○
● ○ ● ● ○   ~   ○ ● ○ ○ ●
● ● ○ ○ ●   ~   ○ ○ ● ● ○
● ● ○ ● ○   ~   ○ ○ ● ○ ●
● ● ● ○ ○   ~   ○ ○ ○ ● ●
```

There are two further ways of lighting the bulbs, namely all the bulbs are on

○ ○ ○ ○ ○

or all the bulbs are off

● ● ● ● ●

This exhausts all the possibilities and shows that in all there are

$$5 + 5 + 10 + 10 + 1 + 1 = 32$$

ways of lighting the bulbs.

Second solution

1) Suppose there is only one bulb. Then the bulb can either be off ● or on ○, i.e., there are two ways of lighting the bulb.

2) Suppose there are two bulbs. The first bulb can be in two states ● or ○. Each of these two states can be combined with the two states of illumination of the second bulb. If the second bulb is off, there are two states

●
  ＞ ● ,
○

while if the second bulb is on, we have

●
  ＞ ○ ,
○

giving a total of 4 ways of lighting the bulbs:

● ●
○ ●
● ○
○ ○

3) Three bulbs. The first two bulbs can be in four states, and each of these 4 states can be combined with either of the two states of the third bulb:

● ●   ● ●
○ ●   ○ ●
● ○   ●   ● ○   ○
○ ○   ○ ○

This gives a total of 8 ways of lighting the bulbs:

● ● ●
○ ● ●
● ○ ●
○ ○ ●
● ● ○
○ ● ○
● ○ ○
○ ○ ○

4) Four bulbs. The first three bulbs can be in 8 states, and each of these states can be combined with either of two states of the fourth bulb, i.e., there are $8 \cdot 2 = 16$ ways of lighting the bulbs.

5) Five bulbs. There are now $16 \cdot 2 = 32$ ways of lighting the bulbs.

The method just described can be modified slightly. For example, another way of describing the transition from two to three bulbs is

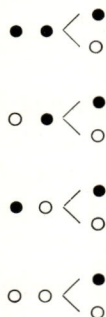

$$● ● \Big\langle \begin{matrix} ● \\ ○ \end{matrix}$$

$$○ ● \Big\langle \begin{matrix} ● \\ ○ \end{matrix}$$

$$● ○ \Big\langle \begin{matrix} ● \\ ○ \end{matrix}$$

$$○ ○ \Big\langle \begin{matrix} ● \\ ○ \end{matrix}$$

In any event, adding another bulb (which can be in either of two states) doubles the number of ways of lighting the bulbs.

45. Let D_n denote the total number of ways of lighting n bulbs. Then $D_j = 2D_{j-1}$, since adding an extra bulb doubles the number of ways of lighting the bulbs (the jth bulb can be off or on).

But obviously $D_1 = 2$. Therefore

$$D_n = 2D_{n-1} = 2^2 D_{n-2} = \cdots = 2^{n-1} D_1 = 2^n \qquad (1)$$

(this is essentially a proof by induction, anticipated in the second solution to Prob. 44). On the other hand, D_n can be calculated by simple enumeration of the possibilities:

0) No bulbs are on. This can occur in C_0^n ways. But obviously there is only one way in which no bulbs are on, regardless of the value of n. Therefore $C_0^n = 1$ for every n.

1) There are C_1^n ways in which one bulb can be on.

2) There are C_2^n ways in which two bulbs can be on.

k) More generally, there are C_k^n ways in which k bulbs can be on.

n) Finally, there are C_n^n ways in which all n bulbs can be on (it is again obvious that $C_n^n = 1$).

It follows that there are a total of

$$D_n = C_0^n + C_1^n + \cdots + C_k^n + \cdots + C_n^n \qquad (2)$$

ways of lighting n bulbs. Comparing (1) and (2), we find that

$$C_0^n + C_1^n + \cdots + C_k^n + \cdots + C_n^n = 2^n,$$

as required. (To make the formula "more elegant," we write C_0^n and C_0^n instead of 1.)

46. Let D_n denote the total number of ways of lighting n traffic lights. Then $D_j = 3D_{j-1}$, since adding an extra traffic light triples the number of ways of lighting the lights (the jth light can be green, yellow or red). But obviously $D_1 = 3$. Therefore

$$D_n = 3D_{n-1} = 3^2 D_{n-2} = \cdots = 3^{n-1} D_1 = 3^n.$$

47. There are 3^k ways of lighting k lights, each of which can be green, yellow or red (see Prob. 46), and 2^{n-k} ways of lighting $n - k$ lights, each of which can be green or red (see Prob. 45). Thus there are a total of $3^k 2^{n-k}$ ways of lighting all the lights,

since for each of the 3^k ways of lighting the group of k lights, there are 2^{n-k} ways of lighting the remaining lights.

48. $26^3 \cdot 10^4 = 175{,}760{,}000$ license plates (there are 26 letters in the alphabet).

49. Suppose there are a total of D_n n-digit numbers containing no zeros or eights. Then $D_j = 8D_{j-1}$, since the extra digit can be 1, 2, 3, 4, 5, 6, 7 or 9. But obviously $D_1 = 8$. Therefore

$$D_n = 8D_{n-1} = 8^2 D_{n-2} = \cdots = 8^{n-1}D_1 = 8^n.$$

In particular, there are

$$D_6 = 8^6 = 262{,}144$$

six-digit numbers containing no zeros or eights.

50. Imagine that the positions of all the teeth in the mouth have been numbered once and for all from 1 to 32. Then to each inhabitant of the city assign a sequence of 32 zeros and ones according to the following rule: The first term is a 1 if there is a tooth in the first position and a 0 if not, the second term is a 1 if there is a tooth in the second position and a 0 if not, and so on. Since different inhabitants are assigned different sequences, the maximum number of inhabitants of the city is simply equal to the total number of such sequences, namely 2^{32} (recall the solutions to Probs. 44 and 45). Incidentally, $2^{32} \approx 4{,}000{,}000{,}000$ which exceeds the present population of the earth.

51. Consider the terms obtained after multiplying out the expression

$$(x - 1)(x - 2) \cdots (x - 100)$$

but before combining terms involving the same power of x. Each such term is the product of 100 factors, and each factor is either the letter x or a number. The first factor is either x or -1, the second is either x or -2, and so on. We are only interested in products in which x appears 99 times and a number appears once. There are 100 such products, since the number can appear in the first place, the second place, and so on, up to the hundredth place. In other words, the products are

$$-1 \cdot \underbrace{x \cdot x \cdots x}_{99 \text{ times}}, \quad x \cdot (-2) \cdot \underbrace{x \cdot x \cdots x}_{99 \text{ times}}, \quad \ldots, \quad \underbrace{x \cdot x \cdots x}_{99 \text{ times}} \cdot (-100).$$

Therefore the coefficient of x^{99} equals

$$-(1 + 2 + \cdots + 100) = -\frac{100 \cdot 101}{2} = -5050$$

(recall Example 1, p. 4).

52. Case 1 (Order is important)

In this case each sum is determined by the first term, which can be any positive integer less than n. Thus there are $n - 1$ ways of writing the sum.

Case 2 (Order is unimportant)

If n is odd, there are only half as many ways as in Case 1, since the sums

$$n = 1 + (n - 1), \quad n = (n - 1) + 1,$$

$$n = 2 + (n - 2), \quad n = (n - 2) + 2,$$

etc., are regarded as the same. Hence the answer is now $\frac{1}{2}(n - 1)$. If n is even, the sum

$$n = \frac{n}{2} + \frac{n}{2}$$

has no "mate," but all the others do. Therefore the answer is now

$$1 + \frac{(n - 1) - 1}{2} = \frac{n}{2}.$$

53. First we find the number of sums where the first term equals 1. In this case, the sum of the second and third terms equals $n - 1$. Clearly there are as many such sums as there are ways of writing the number $n - 1$ as a sum of two terms, i.e.,

$n - 2$ ways as we know from the solution of the preceding problem. Similarly, there are $n - 3$ sums such that the first term equals 2, and so on, with just one sum such that the first term equals $n - 2$. This gives a total of

$$(n - 2) + (n - 3) + \cdots + 2 + 1 = \frac{(n - 2)(n - 1)}{2}$$

ways of writing n as a sum of three positive integers (recall Example 1, p. 4).

Now use the same method to show that the number of ways of writing a positive integer $n \geqslant 4$ as a sum of four positive integers equals

$$\frac{(n - 1)(n - 2)(n - 3)}{6}.$$

Another solution of Problem 53 will be found in the remark on p. 64.

54. If S_n denotes the sum of the numbers in the $(n + 1)$st row of Pascal's triangle, then $S_{n+1} = 2S_n$. In fact, writing the nth and $(n + 1)$st rows as

$$1 \qquad a \qquad b \quad f \qquad g \qquad 1$$

$$0 + 1 \quad 1 + a \quad a + b \quad \cdots \quad f + g \quad g + 1 \quad 1 + 0$$

$$\cdot \; \cdot$$

we find that

$$S_n = (0 + 1) + (1 + a) + (a + b) + \cdots$$
$$+ (f + g) + (g + 1) + (1 + 0).$$

Combining the first terms in each set of parentheses and then the second terms, we have

$$S_n = (0 + 1 + a + \cdots + f + g + 1)$$
$$+ (1 + a + b + \cdots + g + 1 + 0)$$
$$= 2(1 + a + b + \cdots + g + 1) = 2S_{n-1}.$$

But obviously
$$S_1 = 1 + 1 = 2,$$
and hence
$$S_n = 2S_{n-1} = 2^2 S_{n-2} = \cdots = 2^{n-1} S_1 = 2^n,$$

as asserted.

55. The hint to the problem shows that it is indeed possible to put 14 bishops on the board without any two threatening each other. But we must still show that if more than 14 bishops

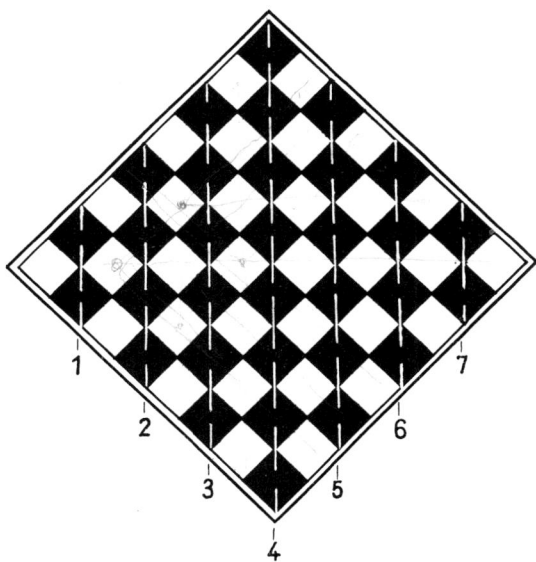

Fig. 8

are put on the board, then two necessarily threaten each other. First we ask how many bishops can be put on the black squares without any two threatening each other. The answer is at least 7 (recall Figure 6), but no more as we know show. To this end, we note that only one bishop can be put on each of the 7 vertical "black diagonals" shown in Figure 8, since two bishops on the same black diagonal will necessarily threaten each other.

There are also 8 horizontal black diagonals (two consisting of a single square) perpendicular to the vertical diagonals shown in Figure 8, but if more than 7 bishops are put on these diagonals, two will necessarily threaten each other. In fact, if a bishop is put on each of the 8 horizontal black diagonals, then the two on the diagonals consisting of a single square would be at opposite corners of the board and hence would necessarily threaten each other. However, there is no need to consider the horizontal black diagonals in Figure 8, since the same answer is obtained by considering just the vertical black diagonals.

A similar argument shows that 7 is the largest number of bishops that can be put on the white diagonals without any two threatening each other. Since a bishop on the white squares cannot threaten a bishop on the black squares, 14 is the largest number of bishops that can be put on the whole board without threatening each other.

Now let W denote the number of ways of putting bishops on white squares, B the number of ways of putting bishops on black squares and N the number of ways of putting bishops on squares of either color, in each case without any two threatening each other. Then clearly

$$W = B, \quad N = BW = B^2,$$

i.e., N is a perfect square.

Remark. It is not hard to solve the more general problem of an $n \times n$ chessboard (i.e., a chessboard n squares on a side) where n is even. Prove that $2n - 2$ bishops can be put on such a board without any two threatening each other and that this number of bishops can be put on in 2^n different ways.

56. One way of giving out the apples and pears is

$$\circ \; \bullet \; \circ \; \bullet \; \bullet,$$

i.e., apples on the first and third days, pears on the second, fourth and fifth days. Another way is

$$\bullet \; \bullet \; \bullet \; \circ \; \circ$$

(pears the first three days, apples the next two days). Thus we must enumerate all arrays made up of two light circles and three dark circles. But this has already been done in the solution of Problem 44, where the answer was found to be $C_2^5 = 10$.

57. Every way of giving out the apples and pears corresponds to an array of k light circles and n dark circles, with days on which apples are given out indicated by light circles and days on which pears are given out indicated by dark circles. Therefore the problem is completely analogous to that of finding all ways of lighting $k + n$ bulbs such that k bulbs are on, a quantity we denote by C_k^{k+n} (recall Prob. 45).

58. $C_2^9 C_3^7$.

59. According to the hint, there is one and only one rook in each file. Moreover, there is obviously one and only one rook in each row. The rook in file a can be placed on any of 8 squares if all 8 rows are previously free. But as a result of putting this rook on the board, one row is no longer free (for example, if the rook is on square a2, no further rooks can be put in row 2). Hence there are now only 7 ways of putting a rook in file b (in the case just considered, it can be placed on squares b1, b3, b4, ..., b8). As a result of putting rooks in files a and b, two rows are occupied, i.e., there are now only 6 free rows left and hence only 6 ways of putting a rook in file c. Similarly, there are 5 ways of putting a rook in file d, 4 ways in file e, 3 ways in file f, 2 ways in file g and only one way in file h.

In other words, there are 8 ways of putting a rook in file a, $8 \cdot 7$ ways of putting rooks in files a and b, $8 \cdot 7 \cdot 6$ ways of putting rooks in files a, b and c, and so on. Hence there are

$$8 \cdot 7 \cdot 6 \cdot 5 \cdot 4 \cdot 3 \cdot 2 \cdot 1 = 8!$$

ways of putting rooks in files a, b, c, ..., h (! denotes the factorial as in Prob. 5).

To check your understanding of this solution, ask yourself in how many ways n rooks can be put on an $n \times n$ chessboard without any two rooks threatening each other. The answer is of course n! Make sure you see why.

We now solve the problem by another method, which gives a different answer (?!). Hence this new solution must contain an error, at least if you believe in the correctness of the first solution. The second solution will be given for an $n \times n$ chessboard from the outset. Suppose n rooks are put on an $n \times n$ board, with no two rooks threatening each other. Then the first rook can be placed on any one of n^2 squares, for example on the square shown in Figure 9a for the case $n = 4$ (the other rooks are not indicated). As in the figure, we shade all squares threatened by this

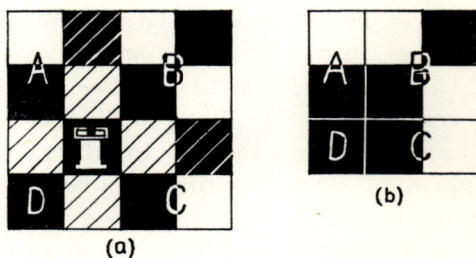

(a)

(b)

Fig. 9

rook, including the square occupied by the rook itself. Then the remaining $n - 1$ rooks must all occupy unshaded squares, since otherwise one of them would threaten the first rook. Next we delete the shaded squares, which together make up a row and a file, leaving behind four pieces A, B, C and D (illustrated in Fig. 9a for the case $n = 4$). By moving these pieces parallel to themselves, they can be assembled into an $(n - 1) \times (n - 1)$ board, as shown in Figure 9b (where $n = 4$ again). No two rooks on the $(n - 1) \times (n - 1)$ board threaten each other, since otherwise they would have threatened each other on the original $n \times n$ board. Moreover, distinct arrangements of rooks on the new board correspond to distinct arrangements of rooks on the original board, and every arrangement of rooks on the new board can be obtained by deleting a row and a file from the original board. As already noted, there are n^2 possible choices of the square ossupied by the first rook. It follows that

$$L_n = n^2 L_{n-1}, \qquad (1)$$

where L_n denotes the total number of arrangements of rooks on an $n \times n$ board such that no two rooks threaten each other. But obviously

$$L_1 = (1!)^2 = 1 \qquad (2)$$

(there is only one way of putting a rook on a board consisting of a single square). Together (1) and (2) imply

$$L_n = n^2 L_{n-1} = n^2 (n - 1)^2 L_{n-2} = \cdots = n^2 (n - 1)^2 \cdots 2^2 L_1^2$$

$$= n^2 (n - 1)^2 \cdots 2^2 \cdot 1^2 = (n!)^2,$$

in seeming contradiction with the first solution. Where is the mistake?

Pay close attention as we now reconcile the two solutions: According to the second solution, there are $(n!)^2$ distinct ways

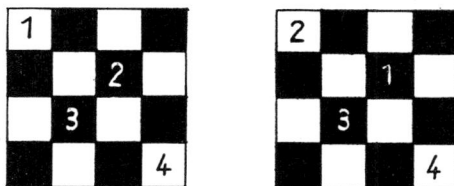

Fig. 10

of putting n rooks on an $n \times n$ board without any two rooks threatening each other, while according to the first solution, there are only $n!$ ways. The second solution would be correct if the rooks were numbered, since then two arrangements like those shown in Figure 10 would be distinct. However, if the rooks are unnumbered, the two arrangements are the same, as in Figure 11, and then the first solution is the correct one.

Fig. 11

Thus we have in fact proved two distinct results:

1) There are $l_n = n!$ arrangements of n indistinguishable rooks on an $n \times n$ chessboard such that no two rooks threaten each other.

2) There are $L_n = (n!)^2$ arrangements of n distinguishable rooks on an $n \times n$ board such that no two rooks threaten each other.

Each of these propositions implies the other, as we now verify for the case $n = 4$ (the proof generalizes at once to the case of arbitrary n). On a 4×4 board, find 4 squares with the property that if 4 rooks are placed on these squares, then no two rooks threaten each other. Label these squares a, b, c and d. Clearly there are l_4 different ways of choosing the squares. Now label the rooks themselves with the numbers 1, 2, 3, and 4, and let P_4 be the number of distinct arrangements of the rooks 1, 2, 3 and 4 on the squares a, b, c and d. Obviously P_4 does not depend on the choice of the squares a, b, c and d. In fact, if A, B, C and D are any other squares with the same property, there are still P_4 distinct arrangements of the rooks 1, 2, 3 and 4 on the squares A, B, C and D. It follows that

$$L_4 = P_4 l_4.$$

But obviously $P_4 = 4!$ (why?), and hence

$$L_4 = 4! \, l_4 = (4!)^2.$$

More generally, if P_n denotes the number of distinct arrangements of n numbered rooks on n fixed squares of a chessboard, then

$$L_n = P_n l_n, \quad P_n = n!,$$

and hence $l_n = n!$ implies $L_n = (n!)^2$ and conversely.

60. Label the n objects 1, 2, ..., n. Then every way of choosing two of the n objects corresponds to a pair of numbers (the numbers of the objects chosen), where obviously the pairs (k, m)

and (m, k) correspond to the same way. Therefore C_2 equals the number of pairs

$$(1,2), (1,3), \ldots, (n - 1, n).$$

The first number in the pair can be chosen in n ways and the second number can be chosen in $n - 1$ ways, giving $n(n - 1)$ ways in all. But the pairs of numbers (k, m) and (m, k) are counted twice, although they represent the same pair of objects. It follows that

$$C_2^n = \frac{n(n - 1)}{2}.$$

Remark. We can now give another solution of Problem 53. Writing a number n as a sum of three terms can be thought of as gathering n beads strung along a wire (as in an abacus) into three groups. Since there are $n - 1$ spaces between the beads, we in effect must choose two of the spaces (where the beads will be separated). But this can be done in

$$C_2^{n-1} = \frac{(n - 2)(n - 1)}{2}$$

ways.

61. $C_2^{28} \cdot 26! = 28!/2$, where, as always, $n!$ denotes the product $1 \cdot 2 \cdots n$.

62. Let P_5^8 denote the number of ways of giving 5 apples, no two of which are alike, to 8 sons. Then P_5 can be calculated by two methods:

1) The number of ways of choosing 5 of the 8 sons (namely those to receive apples) equals C_5^8. But for each such group of sons, there are 5! ways of giving out the apples (the apples are distinguishable). Therefore

$$P_5^8 = C_5^8 \cdot 5!$$

2) The first apple can be given to any of the 8 sons, the second apple can be given to any of the remaining 7 sons, and

so on, up to the fifth apple which can be given to any of 4 sons (the other 3 get no apples). It follows that

$$P_5^8 = 8 \cdot 7 \cdot 6 \cdot 5 \cdot 4 = \frac{8!}{3!}$$

63. The proof is by induction:

a) The result is true for $n = 1$. In fact, the first number in the second row of Pascal's triangle is $1 = C_0^1$ and the second number is $1 = C_1^1$.

b) Suppose the result is true for $n = j$. Then it is also true for $n = j + 1$. This is obvious for $k = 0$ or $\text{k} = j + 1$, since $C_0^{j+1} = 1$, $C_{j+1}^{j+1} = 1$, while the first and last numbers in any row of Pascal's triangle are both 1. Recalling that C_k^j is the number of ways of turning on k out of j bulbs, we now ask how many ways there are of turning on k out of $j + 1$ bulbs. If the $(j + 1)$st bulb is on, there are C_{k-1}^j ways of illuminating the other bulbs (any $k - 1$ of the remaining j bulbs can be turned on). On the other hand, if the $(j + 1)$st bulb is off, there are C_k^j ways of illuminating the other bulbs (any k of the remaining j bulbs can be turned on). It follows that

$$C_k^{j+1} = C_{k-1}^j + C_k^j \tag{1}$$

But C_k^j is the $(k + 1)$st number in the $(j + 1)$st row of Pascal's triangle, by hypothesis, while the next row consists of the numbers

$$1, \ C_0^j + C_1^j, \ C_1^j + C_2^j, ..., C_{j-1}^j + C_j^j, \ 1,$$

by definition (recall Prob. 53). Using (1) we can write this row as

$$1, \ C_1^{j+1}, \ C_2^{j+1}, ..., C_j^{j+1}, \ 1.$$

Therefore the result is true for $n = j + 1$, as asserted, and hence for arbitrary n, by induction.

64. $C_4^8 \cdot 8 \cdot 7 \cdot 6 \cdot 5$.

65. $C_1^3 C_3^{40}$.

66. We begin by finding the number of ways in which 3 objects can be chosen from n objects with regard for order. First we choose any of the n objects, then any of the remaining $n - 1$ objects, and finally any of the $n - 2$ objects which still remain. Thus the three objects can be chosen in $n(n - 1)(n - 2)$ ways if order is important. But this gives every group of objects 6 times if order is unimportant, since the objects A, B and C (say) can be chosen in any of the 6 distinct orders ABC, ACB, BAC, BCA, CAB, CBA. Hence there are

$$C_3^n = \frac{n(n - 1)(n - 2)}{6} = \frac{n(n - 1)(n - 2)}{1 \cdot 2 \cdot 3}$$

ways of choosing 3 objects out of n without regard for order (for the meaning of C_3^n, or more generally C_k^n, see Probs. 45 and 63).

67. By definition, C_k^n is the number of ways of choosing k objects out of n (without regard for order). Generalizing the calculation of C_3^n given in the preceding problem, we find at once that

$$C_k^n = \frac{n(n - 1) \cdots (n - k + 1)}{1 \cdot 2 \cdots k} = \frac{n(n - 1) \cdots (n - k + 1)}{k!},$$

$$(1)$$

where $k = 1, 2, ..., n$ and $k! = 1 \cdot 2 \cdots k$. Formula (1) can be written even more concisely as

$$C_k^n = \frac{n!}{k!(n - k)!} \qquad (2)$$

and holds for $k = 0$ if we define $0! = 1$.

It is easy to prove (2) by induction in n. The formula clearly holds for $n = 1$, since

$$C_0^1 = \frac{1!}{0! \, 1!} = 1, \quad C_1^1 = \frac{1!}{1! \, 0!} = 1.$$

Suppose (2) holds for $n = j$. Then it follows from the formula

$$C_k^{j+1} = C_{k-1}^j + C_k^j$$

proved in the solution to Problem 63 that

$$C_k^{j+1} = \frac{j!}{(k-1)!\,(j-k+1)!} + \frac{j!}{k!\,(j-k)!}$$

$$= \frac{j!}{(k-1)!\,(j-k)!} \left(\frac{1}{j-k+1} + \frac{1}{k} \right)$$

$$= \frac{j!}{(k-1)!\,(j-k)!} \frac{k+(j-k+1)}{k\,(j-k+1)}$$

$$= \frac{j!}{(k-1)!\,(j-k)!} \frac{j+1}{k\,(j-k+1)},$$

where in the second step we use the obvious relations

$$k! = k\,(k-1)!, \quad (j-k+1)! = (j-k+1)\,(j-k)!$$

Therefore

$$C_k^{j+1} = \frac{(j+1)!}{k!\,(j+1-k)!},$$

i.e., (2) holds for $n = j + 1$. Hence, by induction, (2) holds for arbitrary n.

The quantity C_k^n is called *the number of combinations of n objects taken k at a time*, equal to the number of ways of choosing k out of n objects without regard for order. There is also a special notation for the number of ways of choosing k out of n objects with regard for order, namely P_k^n. This notation has already been used in the solution to Problem 62, where we proved that

$$P_5^8 = C_5^8 \cdot 5! = \frac{8!}{3!}$$

More generally, we have

$$P_k^n = C_k^n k! = \frac{n!}{(n-k)!},$$

as you should verify. The quantity P_k^n is called *the number of permutations of n objects taken k at a time*. If $k = n$, the quantity $P_k^n = P_n^n$ is simply called *the number of permutations of n objects*, often denoted by P_n.

68. There are

$$P_6^{88} = \frac{88!}{6!}$$

ways of playing six different consecutive notes. The number of six-note chords equals

$$C_6^{88} = \frac{88!}{6!\,82!}.$$

69. Each of the terms obtained after multiplying out the expression

$$(1 + a)(1 + b)(1 + c)(1 + d)(1 + e)(1 + f)(1 + g)$$

is the product of 7 factors, there being 7 sets of parentheses, where each factor is either a letter or the number 1. Thus we must find the number of products of 7 factors, each of which can be "in two states." In other words, our problem is equivalent to that of finding all ways of lighting 7 bulbs, each of which can be either off or on. The answer is clearly 2^7 (cf. Probs. 44 and 45).

70. We are interested in products in which letters appear three times and the number 1 appears four times (recall the solution to Prob. 69). This is just the number of ways of choosing 3 out of 7 letters, i.e.,

$$C_3^7 = \frac{7 \cdot 6 \cdot 5}{1 \cdot 2 \cdot 3} = 35.$$

More generally, we might multiply out the expression

$$(1 + a_1)(1 + a_2) \cdots (1 + a_n)$$

and ask for the number of terms which are products of k letters. Clearly the answer is now C_k^n by exactly the same reasoning.

71. Each of the terms obtained when the expression

$$(1 + x + y)^{20}$$

is multiplied out without combining terms is a product of 20 factors, where each factor is 1, x or y. Thus we must find the number of products of 20 factors each of which can be "in three states." In other words, our problem is equivalent to that of finding all ways of lighting 20 traffic lights, each of which can be green, yellow or red. The answer is clearly 3^{20} (cf. Prob. 46).

72. The coefficient of x^{17} is

$$\frac{20 \cdot 19 \cdot 18}{2},$$

The coefficient of x^{18} is 0.

73. The expression

$$(1 + a_1)(1 + a_2) \cdots (1 + a_{56}) \qquad (1)$$

reduces to

$$(1 + x)^{56}$$

if we set

$$a_1 = a_2 = \cdots = a_{56} = x. \qquad (2)$$

But according to the solution to Problem 70, multiplying out the expression (1) leads to C_8^{56} terms which are products of 8 letters, and obviously each of these terms becomes x^8 after making the substitution (2). It follows that the coefficient of x^8 in $(1 + x)^{56}$, after multiplying out the expression and combining similar terms, equals C_8^{56}. Similarly, the coefficient of x^{48} equals C_{48}^{56}. The numbers C_8^{56} and C_{48}^{56} are equal, and hence so are the coefficients of x^8 and x^{48} in $(1 + x)^{56}$. In just the same way, the coefficients of x^6 and x^{50} are equal. More generally, the coefficients of x^k and x^{n-k} in the expression $(1 + x)^n$ are equal, since

$$C_k^n = \frac{n!}{k!\,(n - k)!} = \frac{n!}{(n - k)!\,k!} = C_{n-k}^n.$$

74. The proof is almost literally the same as that of Problem 73.

75. The proof is by induction:

a) The formula

$$(1 + x)^n = a_0 + a_1 x + a_2 x^2 + \cdots + a_{n-1} x^{n-1} + a_n x^n \quad (1)$$

holds for $n = 1$, since

$$(1 + x)^1 = 1 + 1 \cdot x$$

and in this case the binomial coefficients obviously coincide with the numbers in the second row of Pascal's triangle.

b) Suppose (1) holds for $n = j - 1$, so that

$$(1 + x)^{j-1} = a_0 + a_1 x + \cdots + a_k x^k + \cdots + a_{j-1} x^{j-1},$$

where $a_0, a_1, \ldots, a_k, \ldots, a_{j-1}$ are the numbers in the jth row of Pascal's triangle. Then (1) also holds for $n = j$, i.e.,

$$(1 + x)^j = b_0 + b_1 + b_1 x + \cdots + b_k x^k + \cdots + b_j x^j, \quad (2)$$

where $b_0, b_1, \ldots, b_k, \ldots, b_j$ are the numbers in the $(j + 1)$st row of Pascal's triangle. To see this, we note that

$$b_0 = a_0 (= 1), b_1 = a_0 + a_1, \ldots, b_k = a_{k-1} + a_k, \ldots,$$

$$b_{j-1} = a_{j-2} + a_{j-1}, b_j = a_{j-1} (= 1),$$

by the definition of Pascal's triangle, so that (2) becomes

$$(1 + x)^j = 1 + (a_0 + a_1) x + \cdots + (a_{k-1} + a_k) x^k$$
$$+ \cdots + (a_{j-2} + a_{j-1}) x^{j-1} + x^j. \quad (3)$$

But (3) is obviously true, since

$$(1 + x)^j = (1 + x)^{j-1}(1 + x)$$
$$= (a_0 + a_1 x + \cdots + a_k x^k + \cdots + a_{j-1} x^{j-1}) (1 + x)$$
$$= a_0 + a_1 x + \cdots + a_k x^k + \cdots + a_{j-1} x^{j-1}$$
$$+ a_0 x + \cdots + a_{k-1} x^k + \cdots + a_{j-2} x^{j-1} + a_{j-1} x^j$$
$$= a_0 + (a_0 + a_1) x + \cdots + (a_{k-1} + a_k) x^k$$
$$+ \cdots + (a_{j-2} + a_{j-1}) x^{j-1} + x^j.$$

It follows by induction that (1) holds for all n.

76. First note that

$$C_0^4 = 1, \ C_1^4 = 4, \ C_2^4 = 6, \ C_3^4 = 4, \ C_4^4 = 1.$$

Therefore

$$\underbrace{100 \cdots 0}_{k \text{ times}} \underbrace{400 \cdots 0}_{k \text{ times}} \underbrace{600 \cdots 0}_{k \text{ times}} \underbrace{400 \cdots 0}_{k \text{ times}} 1$$

$$= C_4^4 10^{4k+4} + C_3^4 10^{3k+3} + C_2^4 10^{2k+2} + C_1^4 10^{k+1} + C_0^4$$

$$= (10^{k+1} + 1)^4$$

(recall Prob. 74), which is clearly a perfect square.

77. Since

$$(a + b)^n = a^n \left(1 + \frac{b}{a}\right)^n,$$

the term involving a^k is

$$C_k^n a^k b^{n-k}.$$

The proof of the binomial theorem is now almost obvious.

78. We first find all terms involving x^k, writing

$$(x + y + z)^n = [x + (y + z)]^n = \cdots + C_k^n x^k (y + z)^{n-k} + \cdots \quad (1)$$

Then in the expression $(y + z)^{n-k}$ we find the term involving y^l:

$$(y + z)^{n-k} = \cdots + C_l^{n-k} y^l z^{n-k-l} + \cdots. \quad (2)$$

It follows from (1) and (2) that the term involving $x^k y^l$ is

$$C_k^n C_l^{n-k} x^k y^l z^{n-k-l}.$$

The explicit formula for $C_k^n C_l^{n-k}$ is

$$C_k^n C_l^{n-k} = \frac{n!}{k!(n-k)!} \frac{(n-k)!}{l!(n-k-l)!} = \frac{n!}{k!\,l!\,(n-k-l)!}.$$

79. Setting $x = 1$ and then $x = -1$ in the binomial theorem

$$(1 + x)^n = C_0^n + C_1^n x + C_2^n x^2 + \cdots + C_n^n x^n,$$

we obtain
$$2^n = C_0^n + C_1^n + C_2^n + \cdots + C_n^n \tag{1}$$
and
$$0 = C_0^n - C_1^n + C_2^n + \cdots + (-1)^n C_n^n. \tag{2}$$

Adding (1) and (2), and then subtracting (2) from (1), we find that
$$C_0^n + C_2^n + C_4^n + \cdots = C_1^n + C_3^n + C_5^n + \cdots = 2^{n-1}.$$

80. The result of multiplying out the expression

$$(1 + x - 3x^2)^{1965}$$

and combining similar terms is some polynomial

$$a_0 + a_1 x + a_2 x^2 + a_3 x^3 + \cdots$$

The sum of the coefficients of this polynomial equals the value of the polynomial for $x = 1$, since

$$a_0 + a_1 \cdot 1 + a_2 \cdot 1^2 + a_3 \cdot 1^3 + \cdots$$
$$= a_0 + a_1 + a_2 + a_3 + \cdots.$$

Hence there is no need to actually carry out complicated algebraic operations. We need only set $x = 1$ in the original expression, obtaining

$$a_0 + a_1 + a_2 + a_3 + \cdots = (1 + 1 - 3 \cdot 1^2)^{1965}$$
$$= (-1)^{1965} = -1.$$

81. There are C_2^n distinct pairs of vertices of the polygon. Joining all these points, we obtain C_2^n segments of which n are sides and the rest diagonals. Hence there are C_2^n diagonals.

82. Being even, the number must end in 0 or 4. If it ends in 0, any of the $4! = 24$ arrangements of the first four digits is allowed. If the number ends in 4, it cannot be allowed to begin with 0, since then it would actually be a four-digit number. This means that the 0 must be in the second, third or fourth position. In each of these three cases there are $3! = 6$ arrangements of the remaining digits 1, 3 and 5. Hence there are a total of

$$24 + 6 + 6 + 6 = 42$$

distinct even five-digit numbers made up of the digits 0, 1, 3, 4 and 5 with no repetitions.

83. $2^6 = 64$.

84. $2^5 = 32$.

85. In the solution to Problem 63, we found that

$$C_{k-1}^n + C_k^n = C_k^{n+1}$$

[change j to n in formula (1), p. 70]. Therefore

$$(C_0^{n+1} + C_1^{n+1}) + C_2^{n+2} + C_3^{n+3} + \cdots + C_k^{n+k}$$
$$= (C_1^{n+2} + C_2^{n+2}) + C_3^{n+3} + \cdots + C_k^{n+k}$$
$$= (C_2^{n+3} + C_3^{n+3}) + \cdots + C_k^{n+k}$$
$$= C_3^{n+4} + \cdots + C_k^{n+k} = \cdots = C_k^{n+k+1},$$

and hence

$$C_1^{n+1} + C_2^{n+1} + C_3^{n+2} + \cdots + C_k^{n+k} = C_k^{n+k+1} - 1,$$

since $C_0^{n+1} = 1$.

86. $C_n^{3n} C_n^{2n} = \dfrac{(3n)!}{(n!)^3}$.

87. The condition of the problem can be written concisely in the form

E	F	G	EF	EG	FG	EFG
6	7	6	2	4	3	1

(E for English, F for French and G for German). Suppose the person who knows all three languages leaves the room. Then among the remaining people, nobody knows more than two languages, and we have the following problem:

E	F	G	EF	EG	FG	EFG
5	6	5	1	3	2	0

Now suppose the three people who know both English and German leave the room. Then the number of people who know other pairs of languages does not change (since nobody in the room knows all three languages):

E	F	G	EF	EG	FG	EFG
2	6	2	1	0	2	0

Finally suppose the two people who know both French and German leave, together with the person who knows both English and French. Then we have

E	F	G	EF	EG	FG	EFG
1	3	0	0	0	0	0

But this problem is trivial, i.e., there is one person in the room who knows only English and three who know only French. Moreover, 7 people in all left the room. Hence the room originally contained 11 people, one of whom knows only English.

88. Suppose the number of subscribers who have borrowed k given books is summed over all possible sets of k books, and let S_k denote this sum. Then

$$S = S_1 - S_2 + S_3 - S_4 + \cdots + (-1)^{n-1}S_n \qquad (1)$$

is the number of people subscribing to the library. To see this, consider a subscriber who has taken out exactly k books, and examine the contribution he makes to each term in the sum (1). There is no loss of generality in assuming that our subscriber has borrowed the first k books. Then he contributes $C_1^k = k$ to the term S_1, since he is one of those who borrowed the first book, one of those who borrowed the second book, etc., up to and including the kth book. Our subscriber contributes C_2^k to S_2, since he is one of those who borrowed every pair of books from among the first k books. Similarly, he contributes C_m^k to S_m if $m \leqslant k$ and 0 to S_m if $m > k$. In other words, he contributes

$$C_1^k - C_2^k + C_3^k - C_4^k + \cdots + (-1)^{k-1}C_k^k$$

to the sum S. But by the solution to Problem 79,

$$-C_0^k + C_1^k - C_2^k + C_3^k + \cdots + (-1)^{k-1}C_k^k = 0,$$

and moreover $C_0^k = 1$. Therefore the contribution of our subscriber to the sum S equals 1. The same argument applies to every subscriber, and hence S equals the number of subscribers to the library, as asserted.

Remark. It is not hard to see that Problem 87 is a special case of Problem 88. On the other hand, the method used to solve Problem 87 can also be used to solve Problem 88.

89. Consider the five sets of telephone numbers of the form

A	B	C
1 2 – – – –	– 1 2 – – –	– – 1 2 – –

D	E
– – – 1 2 –	– – – – 1 2

where the spaces indicate missing digits (for example, B is the set consisting of all numbers with 12 in the second and third places and arbitrary digits elsewhere). Each of the sets A, B, C, D and E contains 10^4 numbers. The sets A and B have no numbers in common, but the sets A and C share 10^2 numbers, i.e., all numbers of the form 1 2 1 2 – –. It is easy to see that there are $6 \cdot 10^2$ numbers which belong to two sets, but only one number, namely 1 2 1 2 1 2, which belongs to three sets. Therefore by the result of the preceding problem, there are a total of

$$5 \cdot 10^4 - 6 \cdot 10^2 + 1 = 49\,401$$

six-digit telephone numbers containing the combination 12.

Test Problems

Group 1 (Sequences and induction)

1. Calculate the sum

$$\frac{1}{1 \cdot 3} + \frac{1}{3 \cdot 5} + \cdots + \frac{1}{(2n - 1)(2n + 1)}.$$

2. Prove that

$$\frac{1^2}{1 \cdot 3} + \frac{2^2}{3 \cdot 5} + \cdots + \frac{n^2}{(2n - 1)(2n + 1)} = \frac{n(n + 1)}{2(2n + 1)}$$

for every positive integer n.

3. Calculate the product

$$\left(1 - \frac{4}{9}\right)\left(1 - \frac{4}{16}\right) \cdots \left(1 - \frac{4}{n^2}\right)$$

where $n \geqslant 3$.

4. Prove that

$$\frac{(n + 1)(n + 2) \cdots (2n - 1) 2n}{1 \cdot 3 \cdot 5 \cdots (2n - 1)} = 2^n.$$

5. Prove that

$$\frac{1}{n + 1} + \frac{1}{n + 2} + \cdots + \frac{1}{2n} > \frac{13}{24}$$

for every integer $n > 1$.

6. For what positive integers does the inequality

$$2^n > 2n + 1$$

hold?

7. Prove that
$$2^{n-1}(a^n + b^n) > (a + b)^n,$$

where $a + b > 0$, $a \neq b$, $n > 1$.

8. Suppose the plane is divided into parts by n circles. Prove that the plane can be colored black and white in such a way that any two adjacent parts have different colors.

9. The sides of an arbitrary convex polygon are shaded from the outside. Then some diagonals are drawn such that no three diagonals intersect in a single point. Each of these diagonals is also shaded on one side, i.e., a narrow shaded strip is drawn along one side of each segment (see Fig. 12). Prove that among the polygons into which the original polygon is divided by the diagonal, at least one is entirely shaded from the outside.

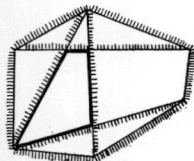
Fig. 12

10. Prove that the number
$$10^{n+1} - 10(n + 1) + n$$

is divisible by 81 for every integer $n \geqslant 0$.

11. Calculate the sum
$$1 \cdot 3 + 3 \cdot 5 + \cdots + n(n + 2).$$

12. The sum of the first m terms of an arithmetic progression equals the sum of the first n terms ($m \neq n$). Prove that the sum of the first $m + n$ terms of the progression equals zero.

13. Calculate the sum
$$6 + 66 + 666 + \cdots + \underbrace{666 \cdots 6}_{n \text{ times}}.$$

14. Does there exist an arithmetic progression with the numbers 1, $\sqrt{2}$ and 3 as terms (not necessarily in this order)?

15. Given a geometric progression whose ratio is an integer other than 0 or -1, prove that the sum of any number of arbitrarily chosen terms cannot equal any term of the progression.

16. Each side of an equilateral triangle is divided into n equal parts, and lines parallel to the sides are drawn through every point of division. As a result, the original triangle is divided into

congruent subtriangles. These subtriangles are then colored, some black and others white, in such a way that every black triangle shares sides with an even number of white triangles, while every white triangle shares sides with an odd number of white triangles. Prove that the subtriangles at the vertices of the original triangle have the same color.

17. Find the fallacy in the following "proof" by induction that all numbers are equal.

Theorem. All numbers are equal.

Proof. A single number equals itself, and hence the theorem is true for $n = 1$. Suppose the theorem is true for $n = k$. Then it is true for $n = k + 1$. In fact, given $k + 1$ numbers, arrange them in some definite order. The first k numbers are equal, by hypothesis, and hence equal the first number. Now eliminate the second number. Then the remaining numbers, which include the first number and the $(k + 1)$st number, are all equal and in particular equal the first number. Therefore all $k + 1$ numbers equal the first number and hence are all equal. The theorem now follows by induction.

Group 2 (Combinatorial problems)

18. In how many distinct ways can the faces of a cube be numbered from 1 to 6?

19. Is 9 or 10 the more probable outcome of tossing a pair of dice?

20. Prove that the numbers in the kth row of Pascal's triangle are odd if and only if $k = 2^n$ ($n = 0, 1, 2, ...$).

21. Which is larger $99^{50} + 100^{50}$ or 101^{50}?

Hint. Write 99 and 101 as $100 - 1$ and $100 + 1$.

22. Prove that the number $11^{100} - 1$ ends in three zeros.

23. How many distinct divisors does the number 10! have?

Hint. $10! = 2^8 3^4 5^2 7^1$.

24. By the *integral part* of x, written $[x]$, is meant the largest integer $\leqslant x$. Prove that

$$[(2 + \sqrt{3})^n]$$

is odd for any positive integer n.

25. a) In how many ways can 19 people be seated at a round table?

b) In how many ways can 19 people be seated at a round table if there are exactly r people between two given people?

26. Find the coefficient of x^{50} in the expression

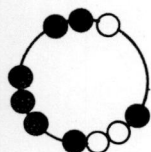

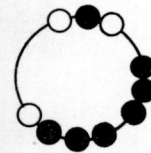

These rattles are identical

$$(1 + x)^{1000} + x(1 + x)^{999} + \cdots + x^{999}(1 + x) + x^{1000}.$$

27. A child's rattle consists of a circular wire on which are strung 3 white beads and 7 red beads. Some rattles look different at first glance but can be made identical by suitably shifting the beads (see Fig. 13). How many different rattles are there?

28. In how many ways can a piece go from square a1 of a chessboard to square h8 (see Figure 5, p. 19) if in each move the piece can go either one square forward or one square to the right?

29. In how many ways can two white rooks and two black rooks be put on a chessboard without a white rook threatening a black rook?

Fig. 13

30. In how many ways can a checker at square a1 of a checkerboard be promoted to a king (if there are no other checkers on the board)?

31. In how many nonself-intersecting ways can one go from A to B in Figure 14? (The motion is along circular and radial routes with stops at the stations indicated by little circles.)

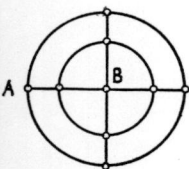

Fig. 14

32. How many ways are there of putting 9 rooks on a $3 \times 3 \times 3$ three-dimensional chessboard without any two rooks threatening each other? (A rook threatens its row, file and vertical column.)

33. The cells of a $2 \times n$ rectangular array are filled in with numbers from 1 to n in such a way that no row or column contains the same number twice. How many such arrays are there?

34. How many five-digit telephone numbers are there containing the combinations 23 and 37?

35. How many n-digit telephone numbers are there containing the combination 12?

36. A father had seven daughters. Every time a daughter got married, each of her older unmarried sisters complained to the father about his violating the custom that the eldest unmarried daughter get married first. After the last daughter was finally married, the father recalled that he had received seven complaints. In how many ways could this have happened?

A more serious statement of this problem goes as follows: By a *permutation* of the numbers from 1 to n we mean these numbers arranged in any order. Any pair of numbers (r, s) in a permutation is called an *inversion* if $r > s$ and r appears to the left of s. How many permutations of the numbers from 1 to n are there containing k inversions? (In the problem of the daughters $n = k = 7$.)

37. If a city has two tall buildings, their spires can be seen in any order (from left to right) if the observer stands at the proper point. The same is true for three tall buildings, provided they do not all lie along some straight line (see Fig. 15). An architect wants to locate seven tall buildings in such a way that their spires can be seen in any order from suitable points. Will he succeed?

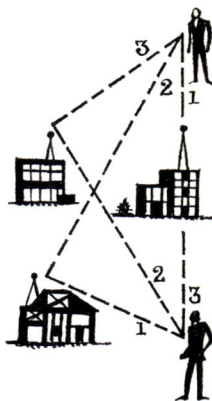

Fig. 15